Emotional Engineering vol. 2

Shuichi Fukuda
Editor

Emotional Engineering vol. 2

 Springer

Editor
Shuichi Fukuda
Stanford University
Stanford, CA
USA

ISBN 978-1-4471-6128-8   ISBN 978-1-4471-4984-2 (eBook)
DOI 10.1007/978-1-4471-4984-2
Springer London Heidelberg New York Dordrecht

Printed on acid-free paper

Springer is part of Springer Science+Business Media (www.springer.com)

Preface

The world we are living in is now expanding quickly. Our world yesterday was closed but now it is wide open. In a closed world, we could make predictions. Designers could foresee how people would behave so they could define missions and could design machines and products. But today situations change very extensively and very frequently. If the world is closed and does not change appreciably, we could make decisions rationally or in other words based on logics of induction and deduction. But it becomes more and more difficult to make decisions and act rationally. All we know are a starting point and a goal, nothing more. If the boundary is clear, then we could apply such a method as reinforced learning which is often applied to navigate a robot in an unstructured environment. But in an open world, boundaries disappear so we have no other ways than to make decisions and take actions by trials and errors.

In a closed world, we could find an optimum but in an open world we cannot even define a global optimum. Thus, we have to come back to the basics of engineering. What is engineering for? Engineering is here to make our dreams come true or to satisfy our expectations. Thus, in an open world, we have to consider how we can satisfy our customers' expectations instead of searching for an engineering optimum. So design is changing quickly to a satisfying design or how we can develop a product which our customers feel good enough. This means performances are more important than functions. We have to get away from our traditional function-based design and move toward satisfaction-focused design. To achieve this goal, we have to know how our customers emotionally reacts to our products and further in order to design a new product, we have to know how they make decisions emotionally.

Yesterday's design was one way from the producer to the customer. It is very much linear. But today's design has to be very reflective. We have to know the feedbacks from our customers. We have to communicate more and more to understand their emotions in order to develop a new product.

Chapters in this volume describe many different aspects of such changes in design and how emotion plays an important role there. This volume contains many sensing issues of emotion, which are crucial in emotional design. Engineering yesterday focused more on machines and we made our efforts to increase their capabilities. It is actuation-focused. But if you read these chapters, you will find out how engineering is changing from actuation-based to sensing-based. Sensing plays a very important role in today's engineering. Yesterday, sensing was carried out to gather information and signals were well defined. But today we do not know what we should sense and how we can sense it; although sensing is very important to understand our customers' emotions, how we can detect emotions is still an open issue today. It is a much complex issue and we have to find our ways how we can sense emotions better. These chapters will provide a good perspective in which direction we should move forward.

I hope this book will help engineers navigate through uncharted waters of tomorrow and surf the big tides of change.

Finally, I would like to thank all the authors from the bottom of my heart for contributing chapters out of their busy schedule and would also like to thank Mr. Anthony Doyle and Ms. Grace Quinn, both at Springer, UK.

Shuichi Fukuda

Contents

Chapter 1
Emotion and Satisficing Engineering

Shuichi Fukuda

Abstract Emotion has been considered to be noises in our traditional engineering because it is situation-dependent and fluctuates very much. But today situations come to change very often and extensively so that it is very difficult to make decisions on a rational basis. If we remember that global optimization is achieved utilizing fluctuations and if we note that AI programs such as the one described by Prolog manages the processing sequence based on verbs, and that verbs are very closely related with our motions, which relates to emotions as etymology indicates. Thus, emotion plays a very important role in our decision making. Although our world is expanding very quickly, the distance and width we can see or what we can experience are limited. Thus Simon's idea of "satisficing" or "satisfied enough" becomes very important. "Satisfied enough" is nothing other than emotion. Therefore, tomorrow's engineering must consider emotion as a very important factor in our product development, because everybody has his or her own perspective of the Open Word, or "My World". He or she makes decisions from the standpoint of My World. Thus, products must be designed and developed to meet expectations which originate from My World.

1.1 Why Emotion Has Been Disregarded in Engineering

Up to now, the world we have lived in was closed with definite boundaries and there were small changes. Therefore, we could make predictions easily because quantitative data were available without any difficulty. Reproducibility has been the primary focus of our traditional engineering so engineering design pursued to be situation-independent. Humans were considered as functional elements just in the same way as machine elements. This is nothing special with engineering. Economics has developed based on such a functional model of economic agents.

S. Fukuda (✉)
Department of Mechanical Engineering, Stanford University, Stanford,
CA 94305, USA
e-mail: shufukuda@cdr.stanford.edu

S. Fukuda (ed.), *Emotional Engineering vol. 2*, DOI: 10.1007/978-1-4471-4984-2_1,
© Springer-Verlag London 2013

But as we enter the twenty first century, situations come to change very often and extensively. To foresee the operating conditions becomes increasingly difficult. It is users who know what is happening now and no one else but only they can respond to changing situations. Therefore, design has changed very quickly from designer-centric to user-centric. Machines not only have to respond flexibly and adaptively to diverse requirements or commands of users, but they also have to sense the changes and provide necessary information. Thus, not only actuation but sensing also becomes very important for a machine.

1.2 Engineering Varies Very Much from Field to Field

There are many different fields in engineering and their ways of thinking vary widely from field to field. Most mechanical engineers regard engineering only in the framework of artifacts. They think they can achieve best quality, if they do their best. They think most parameters are manageable or controllable. For example, if workers' performance is not good, they will replace them with robots and if humidity in a factory is too much high for welding, they will introduce air conditioning. What makes such decisions easy is because most mechanical products are produced in mass so mechanical engineers can make quantitative evaluation without too much difficulty.

But in the case of civil engineering, situations are very much unpredictable in most cases. Their products are used in natural environments and they are produced to order. Unexpected things such as collapse of the ground occur very often. Thus civil engineers are very well aware that some of the constraints are very much hard. Therefore, while mechanical engineers tend to regard most constraints as soft or negotiable, civil engineers know there are truly hard or non-negotiable constraints. Thus, while mechanical engineers make their efforts to achieve the top or desirable quality, civil engineers do their best to secure the bottom or allowable (or acceptable to their customers) quality. In addition to such natural factors, civil engineers have to consider human factors because their products are usually very large and have to be constructed on site.

Shipbuilding engineers are somewhere in between these two. Although ships are constructed in the shipyards, they navigate in natural environments.

There are fields where heuristics are important. Take welding for example. We cannot model welding phenomenon very well. But we can control welding processes by selecting appropriate controllable parameter points in the peripheral region.

1.3 What Is Common Across Fields?

Although the manner of product development differs so widely from field to field, all of them are called "Engineering". Then, what is common across them? Engineering provides us with things not available in nature.

One definition of a human is "Home Faber". We "make" tools. Animals can use tools which are available in nature. But humans are not satisfied with naturally available tools and they make tools to realize their dreams. What differentiates humans from animals is humans can think about the future [1]. The fact we can dream differentiates us from animals. Engineering is an activity of creation. We create our products to make our dreams come true. So how we can meet and satisfy people's expectations is very important and crucial in engineering.

1.4 Twentieth Century was the Age of Dreams

Although it is often said that the twentieth century was an age of products or of material satisfaction, it is not true. Rather, it is an age of dreams. People dreamed about products they would like to have and such products were developed and delivered in response to their expectations. People knew what they wanted and they felt their expectations were being satisfied in the twentieth century.

Apart from such daily life satisfactions, successes such as landing man on the moon convinced people that they were living in an age of dream come true. People were excited to watch their dreams being realized. Thus, the twentieth century is the age of dreams. Many dreams came true. Engineering in the twentieth century was an extension of past experiences. People knew what to expect and their expectations were truly met and satisfied. Therefore, people put trust in engineering.

Such successes were brought about because our world was closed. But as we enter the twenty first century, our world becomes quickly open. The boundaries disappeared and our world keeps on expanding. Yesterday people could make decisions easily because changes were small so they could rely on their experiences. People could expect what will come next. They did not have much difficulty in decision making. In a Closed World, set theory holds and the logics of induction and deduction are effective. Closed World is a world of rationality.

But today never-experienced and unknown products are being developed in large amounts. Therefore people feel uneasy about these new products. How we can help them make appropriate decisions without such anxieties is a pressing issue.

1.5 Twenty first century is the Age of Exploration

In the twentieth century, our world was bounded and closed. So the principle of rationality held true. Set theory was effective and we could apply induction and deduction approaches. We could make appropriate decisions without too much difficulty and act rationally. But as we enter the twenty first century, our world is quickly opening. The boundaries are disappearing and our world is expanding at an unprecedented speed. In such an open world, our traditional approaches are no more effective.

The twenty first century is the age of exploration. We are exploring the New World. In such an Open World, the most important task is to determine our destination. Thus, strategy precedes tactics. It is no longer an age of problem solving. Setting a goal or defining a problem becomes top priority.

If our world is bounded, we can apply such an approach as reinforced learning, even if the path is not known. It enables us to navigate through uncharted waters because there are boundaries.

But the world we are living in today has no more boundaries. The only information we have is about places of departure and arrival. Therefore, only trial and error approaches or reflective approaches such as Abduction, PDSA and Reflective Practice [2–4] can be applied. They are based on Pragmatism. It is none other than the world of "All's well that ends well". The twenty first century is the Age of Pragmatism.

1.6 Emotion and Decision Making

Emotion comes from the Latin words E = Ex (out) + Movere (move). And Motivation and motive come from the same word Movere (move). Thus emotion and motivation constitutes such a cycle as shown in Fig. 1.1.

This is a reflective cycle, same as in Abduction, PDSA and Reflective Practice. Emotion plays an important role in decision making to make another step forward in a feedback system as shown in Fig. 1.2. The closed world is linear. The model is fixed and we know which way we are going. So decision making is very simple,

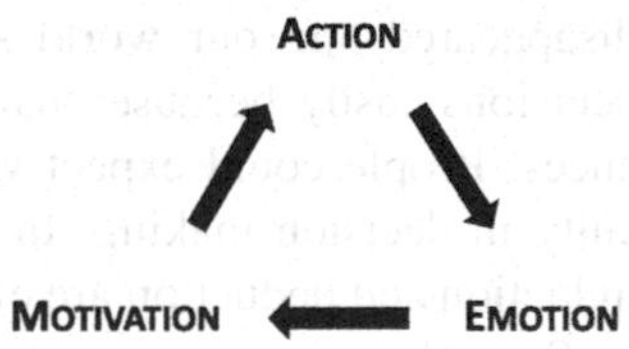

Fig. 1.1 Motivation—action (motion)—emotion cycle

ACTION

MOTIVATION ← EMOTION

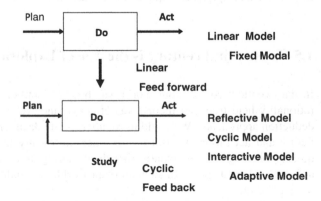

Fig. 1.2 Linear and reflective model

Plan   Act
      Do
                    Linear Model
                       Fixed Modal
            Linear
      Feed forward

Plan   Act
      Do
                    Reflective Model
                    Cyclic Model
      Study   Interactive Model
            Cyclic Adaptive Model
      Feed back

because situations do not change. But with increasing changes of situations, such a linear model is no more effective and we have to convert it to a feedback model. We have to make decisions every time we receive a feedback. And emotion plays an important role there.

1.7 Bounded Rationality

Simon pointed out that our rationality is bounded [5]. If the problem space is not too large, we can make decisions rationally. But if it becomes too large, we rely on emotions. It is because if the problem space is too large, the problem of computational complexity comes in and we cannot solve the problem rationally. Then, we make decisions based upon our emotions.

Another economist Keynes made the same argument. He pointed out that for short term expectations, economic agents make rational decisions. But when it comes to long term expectations, they rely on their confidence, i.e. emotion.

It is interesting to note that Simon is not only an economist but he is also a famous AI researcher. His bounded rationality concept reminds us of the "frame" problem in AI. In the very early stage, AI researchers believed the more knowledge we implement, the wiser a computer will become. So such a word as "knowledge is power" was believed widely and was prevalent. But we soon found out that what we know and learn from our experience are limited so the capabilities of a computer which knowledge is ours are also bounded.

1.8 My World

Thus, the extents we can see and think are bounded, although our world is quickly changing into an Open World. But if we relax our conditions a little bit more by introducing emotion, then we can make decisions beyond the boundaries of strict rationality. Everybody has "My World", where he or she can make appropriate decisions based upon experiences.

Indeed, why do economic agents succeed in long term expectations? Or, why can a brilliant student pass the entrance examination? No matter how much brilliant he or she may be, the problems are new to them. In this sense, there are no difference between brilliant and not so much brilliant students. The greatest difference is that brilliant students are full of confidence and that they take full advantage of their experiences. This is not rational. The way or the solution they choose varies with their personalities and with their experiences.

"My World" is a world where one can make decisions with confidence. The results may not be successful at times, but he or she learns from failures. Thus, "My World" could be described as the world where "learning from failures" or Pragmatism holds.

Thus, "My World" is larger than the world of rationality as shown in Fig. 1.3.

1.9 Emotion and Optimization

Why has emotion been regarded as noises in traditional engineering? It is because it fluctuates so much that it is not easy to treat them as signals. But if we come to think about optimization problems, we become aware that fluctuations play an important role.

Simulated Annealing is a method to optimize globally. By providing fluctuations, we can slip out of local optimum to reach global optimum (Fig. 1.4). In fact, the original annealing provides fluctuations to material elements and ensure greater strength by distributing material elements more homogenously.

And Astumian found out that in molecular motors, Brownian movement plays an important role [6, 7]. Therefore, fluctuations play an important role in our actions and in decision making.

Therefore, emotion may be added to us to optimize our decisions globally.

Fig. 1.3 World of rationality, my world and open world

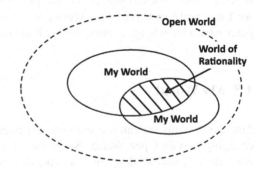

Fig. 1.4 Local optimum (*white*) to global optimum (*black*) using fluctuation (Simulated Annealing)

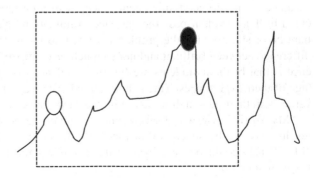

1.10 Reason and Emotion

Damasio pointed out reason and emotion are very closely related and insists reason and emotion cannot be separated [8]. This is supported by many brain researchers these days. If we consider remarkable progresses of brain science today, we may regard "My World" as a world of extended rationality. Traditional world of rationality is that of Simon's and beyond these boundaries there is another region where emotion plays an important role in decision making (Fig. 1.5). Since our field of views and our experience are limited, this extended world is still bounded and "My World" varies from person to person depending upon how far a person can see and what experience he or she may have. Simon's world of rationality is common to all of us, but this extended region varies very extensively from person to person.

1.11 AI, Decision Making and Emotion

One of the approaches in AI is first order predicate logic. Let us take programming language, Prolog (Programming in Logic), for example. Our knowledge is represented as a network with edges representing verbs and nodes representing Subjects or Objects [9]. Thus, our knowledge in the form of S–V–O can be described as a network.

If we remember that in quality function deployment (QFD), functions are described by verbs, we will immediately understand that QFD can be described in the form of a network using such a verb-based logic.

The English language use different verbs for different situations. So the situations can be easily implemented if we use English. In AI programming, information is processed based upon pattern matching. Pattern driven processing is nothing other than decision making because it decides which step will be the next.

Fig. 1.5 World of rationality and my world

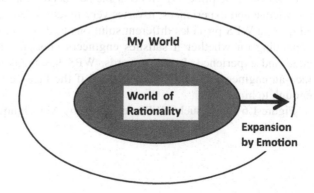

As emotion is very closely associated with our motions or verbs, this demonstrates what an important role emotion plays in our decision making.

1.12 Satisficing vs. Optimizing

Simon proposed Satisficing Theory [10, 11]. Satisficing is the word coined by putting "satisfy" and "sufficing" together. Thus, it means "Satisfied enough". Simon insisted that since our world of rationality is bounded, we cannot optimize our decisions rationality. What we can do is to satisfy our expectations to the degree that we feel "enough". This is very much emotional. He did not use such a word but what he is insisting is the importance of emotional satisfaction.

To describe it in another way, this is to define "My World" appropriately." My World" is the world bounded by how far a person can see and how confidently he or she can make decisions. Simon's "Satisficing" could be interpreted as another definition of "My World".

Up to now, we have been developing products along the line of optimization, but if we consider that our world is very quickly expanding with many frequent and extensive changes, such optimization approach will be no longer feasible and is unrealistic. Such optimization is local. We should move toward global optimization which is "satisficing engineering".

1.13 Illustrative Example

One typical example of such emotional optimization can be found in welding. In welding, Welding Procedure Specification (WPS) is required to carry out welding for important products. But welding is an assembly of many different engineering knowledge and experiences so that there is no unique best optimal solution. WPS is developed with many different inputs from engineers with various disciplines. As WPS is such an accumulation of different sources of knowledge and experiences, it varies very much by who implemented the knowledge. And WPS provides different solutions to different inputs. WPS is finalized depending on whether it satisfies engineers enough who implemented knowledge and experience. In other words, WPS has many faces. No matter which face an engineer may look at, it is good if the face he is looking at looks good enough to him.

Figure 1.6 shows such a WPS developed by AI technique [12].

Fig. 1.6 Welding procedure
specification (*WPS*)

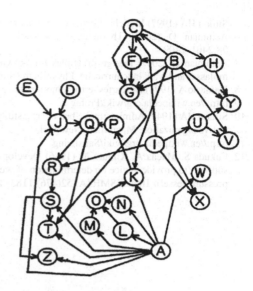

```
A= MATERIAL-TYPE
B= PLATE-THICKNESS
C= WELDING-METHOD
D= HUMIDITY-IN-WELDING-ENVIRONMENT
E= WELDING-MATERIAL
F= GROOVE-TYPE
G= LAYER
H= NUMBER-OF-LAYER
I= HEAT-INPUT
J= HYDROGEN-VALUE
K= PCM
L= CHEMICAL-COMPOSITION-C
M= CHEMICAL-COMPOSITION-MN
N= CHEMICAL-COMPOSITION-SI
O= CHEMICAL-COMPOSITION-SUM
P= PCW
Q= PREHEATING-TEMPERATURE
R= COOLING-TIME
S= METAL-STRUCTURE
T= TOUGHNESS-IN-WELDED-JOINTS
U= EVALUATE-DEFORMATION
V= ANGULAR-DISTORTION
W= TENSILE-RESIDUAL-STRESS-WIDTH
X= MAXIMUM-RESIDUAL-STRESS-VALUE
Y= TRANSVERSE-SHRINKAGE
Z= HARDNESS-IN-WELDED-JOINTS
```

References

1. What will our consciousness about the future, which is specific to human bring to us?
   Applied Brain Science Symposium, Institute of Science of Survival (2010) (in Japanese)
2. http://plato.stanford.edu/entries/abduction/
3. Best M, Neuhauser D (2006) Walter A. Shewhart, 1924, and the Hawthorne factory. Qual Saf
   Health Care 15:142–143
4. Schon DA (1984) The reflective practitioner: how professionals think in action. Basic Books,
   London

5. Simon HA (1997) Models of bounded rationality. MIT Press, Cambridge
6. Astumian D (2010) Thermodynamics and kinetics of molecular motors. Biophys J 98:2401–2409
7. Astumian D (2007) Design principles for brownian molecular machines: how to swim in molasses and walk in a hurricane. Phys Chem Chem Phys 9(37):5067–5083
8. Damasio A (1994) Descartes' error: emotion, reason and the human brain. Putnam, New York
9. http://en.wikipedia.org/wiki/Prolog
10. Simon HA (1947) Administrative behavior: a study of decision-making processes in administrative organization. Macmillan, New York
11. http://en.wikipedia.org/wiki/Satisficing
12. Fukuda S, Maeda A, Kimura M (1986) Development of an expert system for weld design support (to provide advice on determination of weld conditions to prevent weld cracking in a pressure vessel). Trans. JSME (A) 52(476):1183–1190 (in Japanese)

Chapter 2
Emotion and Innovation

Shuichi Fukuda

Abstract The word "innovation" is attracting wide attention these days. It is pointed out in this chapter that quality becomes increasingly difficult to be recognized with its improvement. Quality in its original meaning means segments. Therefore innovation and pursuit of quality concurs on the point unless new sectors are explored, customers do not feel their expectations are met and they will not be satisfied. In the old days, product lifecycle were much longer so people could make decisions whether to buy or not based upon their use experience or upon observations of foregoing customers. But today product lifecycles are getting shorter and shorter so people have to make decisions as soon as new products come up on the market. But such new products are not only experienced before, but also developed with new technologies they do not know. So people feel very much uneasy when they buy innovative products. How we can reduce their anxieties is very much important for our innovative products to be accepted by them. Therefore, true innovation is more emotional than technical. How we can make our innovation more emotional and acceptable is discussed with illustrative cases.

2.1 Why Innovation is Important

Everybody talks about innovation these days. Why is innovation so important? Today changes are so frequent and extensive and the boundaries of our world are quick disappearing so our world becomes an Open World. In the days of a Closed World where situations did not change appreciably and its boundaries were fixed, we could apply set theory and induction and deduction logics (Fig. 2.1).

And as the world is small, there was only one peak so that we did not have to make decisions which peak to climb. Our efforts were fully rewarded in these days. We did not have to think about innovation because Christensen's sustaining innovation [1] was the only option we could take. But as our world expands, many other peaks appear and we have to make decisions which peak to climb (Fig. 2.2).

S. Fukuda (✉)
Department of Mechanical Engineering, Stanford University, Stanford, CA 94305, USA
e-mail: shufukuda@cdr.stanford.edu

S. Fukuda (ed.), *Emotional Engineering vol. 2*, DOI: 10.1007/978-1-4471-4984-2_2,   11
© Springer-Verlag London 2013

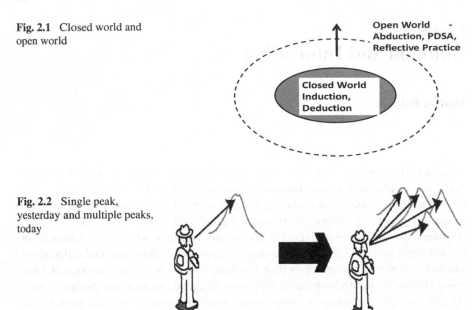

Fig. 2.1 Closed world and open world

Fig. 2.2 Single peak, yesterday and multiple peaks, today

Thus setting a goal or defining a problem becomes more important than problem solving. We have to change our way of thinking from tactics to strategy. And in an Open World, what we have to do is to explore. So trials and errors are prerequisites and we have used such reflective approaches as Abduction, PDSA and Reflective Practice [2–4] (Fig. 2.3).

Christensen pointed out [1] that there are two innovations. Our traditional innovation is sustaining innovation where people make efforts to move forward on the same track. He pointed out, taking hard disk drives as an example that while such efforts are being made; new markets are being created off the track. Such new market emerges from customers' expectations. When situations change very extensively, the producer tends to overlook such emerging markets and they fail to respond to such expectations. Thus, he called such an innovation "disruptive".

Yesterday changes were small. Mathematically speaking, the change curves were smooth and continuous. Thus, they are differentiable. So we could make predictions. Today, changes are frequent and extensive. What makes predictions very difficult is because the change curves are not smooth. They are angular so they are not differentiable. Therefore, it becomes almost impossible for the producer to predict the future (Fig. 2.4).

"Disruptive" is the name which comes from the viewpoint of market. It would be better to call such an innovation "evolving innovation" from the viewpoint of industry or the producer, because such situations necessitates the producer to adapt to changes of customers' expectations.

Innovation becomes important from another reason. We should remember that the higher the quality of a product becomes, the more it is difficult for a customer

Fig. 2.3 Reflective cycle
(Shewhart cycle)

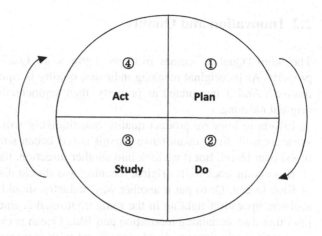

Fig. 2.4 Changes of
yesterday and today

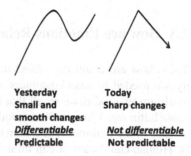

Yesterday
Small and
smooth changes
Differentiable
Predictable

Today
Sharp changes
Not differentiable
Not predictable

to realize its improvement. Weber-Fechner pointed out that we need an increment proportional to its level of stimulus to perceive the difference of a level.

$$\Delta S/S = \text{Constant}$$

$$\Delta S = \text{increment of the stimulus,} \quad S = \text{level of Stimulus}$$

If a man with a small voice raises his voice a little, people will immediately understand he raised his voice. But if a man with a big voice raises his voice a little, people would not understand. Thus, although industry is making tremendous efforts today to improve quality, customers cannot recognize the improvements. To let them recognize, just noticeable difference (JND) [5] is enough. But it is very difficult to know how much JND is.

If we explore the new market, people understand immediately the difference. This is what Kim and Mauborgne [6] pointed out in their book "Blue Ocean Strategy". If is much more effective and rewarding to look for a blue ocean than fighting blood over blood on the red ocean.

2.2 Innovation and Quality

The word "Quality" comes from the Latin word Qualitas, meaning quality or property. As its original meaning indicates, quality is "qualitative" and not "quantitative". And if interpreted as property, then segmentation is much closer to its original meaning.

Efforts to improve product quality quantitatively will not be rewarded in this sense as well. Such quantitative pursuit is red ocean strategy. Battles are fought blood over blood. But if we look into another direction, there is a Blue Ocean.

If we come back to its original meaning, we should discuss quality in the sense of Blue Ocean. Or to put it another way, industry should pursue another business segment instead of fighting in the same traditional segments. Red Ocean is nothing other than sustaining innovation and Blue Ocean is evolving innovation. Thus, pursuit of quality is very closely associated with innovation in the sense that both are activities to explore new segments.

2.3 How are Emotions Related to Innovation and Quality?

Then, how are emotions related to innovation and quality? Innovation and quality are needed to meet customers' expectations. Expectation is emotion as can be easily understood if we recall that the word motivation or motive comes from the same Latin word "movere" as emotion does.

One of the definitions of a human, "Homo Faber", best describes what a human is. Human can dream or can think about the future. Animals cannot. Animals can use tools available in nature, but humans are not satisfied with such naturally available tools. They make tools to realize their dreams. This is emotion. Engineering is an activity of creating artifacts that are not available in nature. It is an activity to realize our dreams. So it may not be too much to say that engineering itself is an emotional activity. Innovation and pursuit of quality comes from our basic human nature and it is very much related to our emotions.

Thus, innovation becomes true innovation when it meets the expectations of customers and when they feel they have new innovative products. Unless they do not feel that way, that is not an innovation, no matter how they are innovated technically.

Thus innovation lies in the hearts of customers. The same argument holds for quality. What is important in quality management is not quality improvement, but quality satisfaction or meeting quality expectations. If customers feel the quality meets their expectations, then that is true quality.

Thus, although quality also lies in the hearts of customers, current discussions about innovation and quality are too much technical focused. We should give more thoughts and attentions to emotional aspects. If a customer feels a product is innovative or good enough, then our job is perfect, no matter whether it is not too much innovative or is very sophisticated in technical quality. True quality is very much subjective and emotional.

2.4 Reverse Innovation

Today, another innovation, i.e., reverse innovation is attracting wide attention. But it is another form of evolving or disruptive innovation. Take GE Healthcare for example. They had to adapt their cardiograph to Indian market because space, etc. are limited there. They had to develop smaller and simpler ones. They did not realize that there were great demands for such small and simple equipments back in their home country. Until then, they were fighting along the line of Red Ocean. But their business in India opened their eyes to the new "Blue Ocean" market in the States. Although it is called "reverse", this is nothing other than another Blue Ocean. It so happened that a developing country business made them aware of such a business opportunity.

Much simpler case can be found in Japan. Such a trailer as shown in Fig. 2.5 is called "rear car".

Although there were many "rear car" makers after the war, most of them shut down their business and only several manufactures survived. One small company, Muramatsu Sharyo was asked by an African Embassy to produce them in order to reduce the burden of women carrying waters to their homes from the well. This story reminded Yamato Unyu, Japanese delivery company, of their usefulness. They do not have parking problems. They can negotiate narrow roads. Now rear-cars as shown in Fig. 2.6 are seen every day, everywhere in Japan. Muramatsu realized that rear cars could be used in a different manner. This success led Muramatsu to develop mobile wagons for flower shops. Most of flower shops in Japanese hotels have to vacate their space after business hours. Rear cars met their expectations.

2.5 Simplicity is the Soul of Innovation

So a lesson we learned is we do not have to make our products complex or sophisticated. Rather, simplicity is the soul of innovation. This is demonstrated by iPhone.

Fig. 2.5 "Rear Car" in Japan

Fig. 2.6 "Rear Car" for delivery

Until the emergence of iPhone, most cell phone makers paid every effort to attach more functions. They believed more functions would satisfy customers better. But it did not work that way. I asked many young students who are studying IT whether they are satisfied with cell phones. To my surprise, most of them told me the way cell phones are being developed at that time made them irritating and led to dissatisfaction. This is very much the opposite from what cell phone developers expected.

Students told me that they would like to master new functions as soon as they are added. But too much complexity made it very difficult to do it in a short time, and more often than not it called for extensive knowledge to really put them into use. So cell phones irritated young students because their expectations are not met. Cell phone makers have been making efforts to discourage users.

iPhone, on the other side, is very much different. The basic mechanism is simple and a user adds apps as he or she wishes or as he or she needs. So it works exactly the way a customer expects. So iPhone was accepted by market very quickly and widely.

How important it is to meet the customers' expectations is already pointed out by Norman [7]. He pointed out yesterday people put more trust in machines because they were simple and operated as people expected. But today they become too much complicated and they do not work as people expect. So people are quickly losing trust. iPhone demonstrates how simplicity is important.

2.6 Multiple Systems for One Human Need

The current framework of industry drags the histories of inventions. They are not organized to correspond to human needs. Let us take transportation for example. Cars, rails, ships and airplanes were invented to solve a very specific technical issue and developed independently.

Airplane was invented to satisfy our desire to fly. At the time of invention, the challenging issue was to fly in the air. That was the immediate objective. But we forgot why we hoped to fly like a bird. Our final goal was to travel safely and comfortable, no matter what situations come up.

In fact, if we look at birds, they fly because they have to travel a long distance. Indeed they fly, but they also walk and swim. If we come back to our basic desire, all transportation industries should be integrated and united into one.

Our industry development has been too much technically driven and we forgot to get down to our basic human desire. If we look industries from this perspective, we would know there are many such cases where integration of industries are called for.

It should be pointed out that such integration leads to a great amount of reduction of cost and energy consumption and it also increases productivity to a great deal. Again "simplicity is the soul of innovation" must be remembered.

Such reorganization of industries may be called "Emotional Reorganization", because industries are reorganized for emotional satisfaction of customers, and not for technical interests. And it should also be added that such integrated transportation system is really called for in such a big country as Brazil. You could fly to a distant place, but if a car is not available there, you cannot go any further.

2.7 Social Innovation

Integrated innovation is needed not just for meeting customer's expectations, but for sustainable development as well. To save energy, we have to integrate our products and industries. Many different products which have been developed independently up to now will be integrated. Such integrations calls for emotional innovation in addition to technical one. We have to decide what kind of a society we want in the future.

New energy management system, for example, is now changing designs of housing, transportation, etc. to much integrated design. Product based designs are now changing into a holistic system design or the design of society. Thus, we will be designing our lifestyle in the future.

2.8 Process Innovation

Maslow pointed out that the highest human need is self actualization (Fig. 2.7).

The main spring and the core of all human activities are challenges. We would like to challenge to new situations. It is deeply associated with evolutionary innovation. The highest human need is our desire to innovate ourselves. This is emotional innovation.

Fig. 2.7 Maslow's hierarchy
of human needs

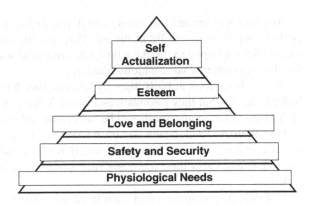

If we could develop such a system where our customers can enjoy producing and repairing their products, then their desire for self actualization will be satisfied. To get our customers involved in design, manufacturing and repair, our design and production system needs an extensive change. Such new DIY system is very much of a challenging issue and certainly calls for technical innovations, but it also brings about emotional innovation.

If we are successful in developing such a new framework of DIY, our customers would feel more attached to such self produced products because their time and efforts have been spent upon them. This is the endowment effect.

When people talk about innovation, they focus on new developments. But if people come to use our products longer, repair will come up as another target for innovation. Current product developments focus on one time value. Values are discussed and evaluated at the point of delivery. But there is lifetime value. Extension of a product lifecycle will increase lifetime value.

Repair is not the same as maintenance. Maintenance is to bring the degrading functions back to its original design level. Repair is to keep products in the best working conditions. The functions or materials degrade, but if they work well in the operating conditions, our customers are happy.

This holds true with humans. What doctors define as good health and our feeling of being healthy are different. Doctor's health is an idealized heath or health as a mass. But our health is very much personal. If we can live our life happily and comfortably, we feel we are healthy.

Repair and remedy are basically the same. They require great amount of knowledge and experience. But if customers can perform such repair jobs, they would be happier, because not only they can enjoy the best working products but also their desire for challenge will be satisfied.

2.9 Emotional Innovation

Since innovation is more emotional than technical, it is important to consider innovation from emotional viewpoint. To illustrate the points, two cases are taken up.

2.9.1 The Product Space: Permeating Innovation

The success of Korea in global market is well known. Korea used to be poorer than Chili because she is small and does not have any appreciable resources. But while Korea succeeded in global market today, Chili still remains the same. This is because Chili is rich with natural resources so Chili did nothing more than just to export them. That way, Chili could easily survive. But Korea, being poor in natural resources, no Korean thinks Korea as a market. Most of them look for the global market from the first.

Hidalgo clarified why Korea succeeded [8]. What Korea took as their strategy is to explore the market very close to the current one using the current available technologies. They studied what products are close to the ones they are producing today. Then they produced such neighboring products.

I would like to call such an innovation "Permeating Innovation". This is one type of evolving innovations. Such permeating innovations not only reduce product development time, cost and resources, but what is more important is it takes away the anxiety of customers.

Today many new products are being developed in much shorter time. Most of these products use new technologies customers never experienced. Yesterday, it took more years to develop a product and their lifecycles were long. About 30 % of foregoing customers tried these new products. So the rest 70 % customers could observe how the product would work and they could decide whether to buy or not based upon the observations. They could buy these new products with confidence.

But today the product lifecycles are getting shorter and shorter. So every customer must make a decision whether to buy or not once a new product comes up on the market. He or she has never used such a product before and what makes them anxious is most of these new products are developed with new technologies he or she has never experienced or never heard of. So their anxiety is increasing. The Korean strategy of "Permeating Innovation" greatly reduces such concerns.

2.9.2 Segway and P.U.M.A

Dean Kamen developed Segway or personal transport (Fig. 2.8).

It was originally developed as a wheelchair but the disapproval of the idea of a new wheelchair by a governmental institution led him to develop this innovative vehicle. What is innovative is not the technology. The idea of an inverted pendulum is nothing new. What is important about Segway is Dean created a new lifestyle. What he developed is not a new product, but a new lifestyle.

Dean's another innovation, P.U.M.A, Personal Urban Mobility and Accessibility, developed it with GM better demonstrates that he is developing a lifestyle. If we compare a wheel chair and P.U.M.A., we would immediately know that there is no difference between a wheelchair and personal urban transport. In fact, he developed P.U.M.A. based on his idea of a wheelchair (Fig. 2.9).

Fig. 2.8 Segway, personal urban transport

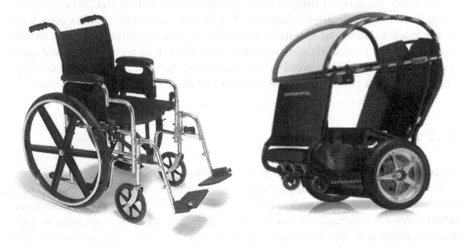

Fig. 2.9 Wheelchair and P.U.M.A

If we could develop such a transport with the same platform but with different optional systems on top of it, then there will be no distinction between a wheelchair and personal urban transport. The current distinction will lose its meaning.

Then, disabled can say that the only difference is their lifestyle. Some may have more mobility and accessibility. But it so happens they do not. Thus not only such a transport would provide pride and joy to everybody, but also would increase productivity to a great extent and reduce cost and energy. What is most important is people can enjoy innovation.

References

1. Christensen CM (2011) The Innovator's Dilemma: the revolutionary book that will change the way you do business. HarperBusiness Reprint
2. http://plato.stanford.edu/entries/abduction
3. Schon DA (1984) The reflective practitioner: how professionals think in action. Basic Books
4. Fukuda S (2011) Emotional engineering. Springer, New York
5. http://en.wikipedia.org/wiki\Just-noticeable_difference
6. Kim WC, Mauborgne R (2005) Blue ocean strategy. Harvard Business Review Press
7. Norman DA (2003) Emotional design: why we love (or hate) everyday things. Basic Books
8. Hidalgo CA, Klinger B, Barabasi AL, Hausman R (2007) The product space conditions the development of nations. Science 317:485

Chapter 3
Touch Feelings and Sensor for Measuring the Touch Feeling

Mami Tanaka

Abstract Touch feelings and tactile sense play very important role in our daily life. However, the mechanism has not clarified yet. In this chapter, the receptor in human skin and haptics that is motion of hand/finger will be introduced. And the sensory experiments and measurement were carried out in order to clarify the mechanism of rough and soft feelings which are fundamental touch feelings and the developed sensor for measuring tactile sensation for fabrics and the palpation sensor for measuring prostatic glands will be introduced.

3.1 Introduction

Touch is one of five senses, and it can feel mechanical stimuli (pressure, vibration, heat, cold, pain, etc.) through the skin. Skin is the largest sense organ in the human body and the area is about 1.8 m^2 in the average adult [1]. Unborn babies have the skin function from the 9th week after conception and babies use their hands and mouth in order to obtain the many information of outside world.

Touch is active and passive and it is ruled under the law of action and reaction of skin and objects, which is called "third law of motion". This point is very unique and tactile sense is unlike other senses for this point. Therefore, it is difficult to clarify the mechanisms of tactile sense and the feeling of touch. The mechanisms of vision and hearing have been already clarified and the principals contribute the development of the glasses and hearing aid in order to assist the vision and hearing sense, respectively. It is very important to clarify the mechanism of the touch sense, like these.

M. Tanaka (✉)
Department of Biomedical Engineering, Tohoku University, Sendai 980–8579, Japan
e-mail: mami@rose.mech.tohoku.ac.jp

S. Fukuda (ed.), *Emotional Engineering vol. 2*, DOI: 10.1007/978-1-4471-4984-2_3,
© Springer-Verlag London 2013

Table 3.1 Sensory receptor and modality and categorization in human skin [1]

Receptor	Modality	Category
Meissner's corpuscle	Touch flutter	FAI
Pacinian corpuscle	Touch vibration	FAII
Merkel's discs	Touch pressure	SAI
Ruffini endings	Touch pressure	SAII
Free nerve endings	Temperature and pain	

3.2 Sensory Receptor in Human Skin

Table 3.1 shows the sensory receptors and modality in human skin [1]. Free nerve endings react for temperature and pain and they are different from the others. The other receptors react to mechanical stimuli and are classified depending on the reaction speed, fast adaptive (FA) and slow adaptive (SA). In addition the receptors are classified depending on the size of area, the area of II means larger than that of I.

As the FA sensory receptors, there are two kinds, Pacinian corpuscle and meissner's corpuscle. Concerning with these FA sensory receptors, they have higher sensitivity frequency ranges. When the receptors receive the sinusoidal wave stimulus, Pacinian corpuscle can react under 1 μm threshold at 250–300 Hz, and Meissner's can react under 10 μm at 30–40 Hz. It is interesting the receptors have different higher frequency ranges.

3.3 Search for the Mechanism About Roughness and Softness

There are various fields to use touch feelings and sensation, for example, to make something in industry, palpation in medical and welfare fields.

In industry, the sensory tests usually are done to evaluate the many things, but a huge number of subjects are needed to obtain the accurate touch sensation. Therefore, the training of the expert for measuring touch feelings is important. Palpation has important role for a clinician/doctor in diagnoses. They assess smoothness, roughness, and softness of an area of patient and/or find the abnormal point such as hard spot by palpation. However, palpation using human's fingers is said to be ambiguous, subjective and much affected by their experience. From these points, it is not easy to share the information of a same diagnosis.

Various kinds of information are obtained as the touch feeling and the roughness and softness are fundamental touch feelings. The physical value of the roughness can be measured by surface roughness measuring instrument as the amplitude of the surface asperity, but the measurement object is limited such as metallic materials and the relationship between the obtained physical value and the touch feeling has not been clarified. In addition, there is not only the hard one like the metal but also various one in the world where we live. These make more difficult the clarification of the mechanism that human feel rough.

The softness of the object is also measured as the stiffness and Young's modulus by hardness tester, but the relation between the values and touch feeling has not been clarified. Therefore, the clarification of the mechanism becomes more difficult.

In order to search the trigger of the mechanism, we investigated the relations between amplitude and frequency information, and smoothness and roughness.

3.4 Tactile Display for the Roughness Tests

A simple tactile display that subjects can touch and feel various degree of smoothness is developed. Bimorph cell is used as an actuator to generate vibratory stimulus. And the display can adjust the frequency of vibratory stimulus. Through two experiments, the relationships between the frequency distribution of vibratory stimulus and smoothness feelings are investigated.

After this, "roughness" and "smoothness" do not mean physical roughness and smoothness of an object, but mean sensuous roughness and smoothness which human feels using their tactile sense.

A simple tactile display using bimorph cells is fabricated. The tactile display is shown in Fig. 3.1. The display consists of an actuator on the display and it is vibrated using bimorph cells with steady frequency and amplitude. Three input waves of the bimorph cells are generated using GNU Octave which is numerical computation software. The generated waves are transmitted to the amplifier and the amplified outputs were applied to the actuators. In experiments, subjects put their forefingers of dominant hand on the actuators and evaluate what they feel. Figures 3.2 and 3.3 show the size of the display and experimental scene, respectively.

3.5 Sensory Tests for Roughness

Two sensory tests about roughness/smoothness are conducted using the tactile display. In the first experiment, the relations between the vibration frequency and "Kansei" keywords of roughness feeling were investigated and in the second

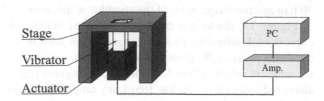

Fig. 3.1 System of tactile display

Stage
Vibrator
Actuator

PC
Amp.

Fig. 3.2 Size of the display

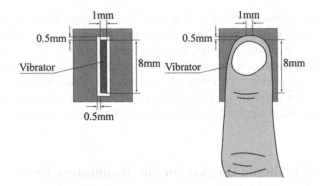

Fig. 3.3 Scene of the tests

experiment the relations between vibration frequency and degrees of roughness feeling were investigated.

In order to find the relations between the vibration frequency and "Kansei" words of roughness feeling, the various vibration stimuli are displayed to the subjects. In the experiment, the subjects touched the tactile display and selected one keyword from six Kansei keywords to answer how they felt it.

Six keywords are as follows. "A: not felt", "B: snaky", "C: uneven", "D: rough", "E: fine" and "F: vibration". The keywords were selected through the preliminary experiment. The subjects are six men and they are 21–32 years old. The tactile display was presented sinusoidal wave vibration stimulus of the amplitude 30 μm and the frequencies of the stimulus were changed from 1 to 250 Hz.

Figure 3.4 shows the ratio of number of subjects that use each keyword to express each stimulus. From this result, it is seen that each Kansei keyword has the corresponding frequency distribution. At the lowest frequency range, most of the subjects selected "not felt", and at the highest frequency range many subjects answered "vibration". At low frequency range, the subjects answered "snaky" and

Fig. 3.4 Ratio of number of subjects that use each keywords to express each stimuli. *A*: "not felt", *B*: "snaky", *C*: "uneven", *D* "rough", *E* "fine" and *F* "vibration"

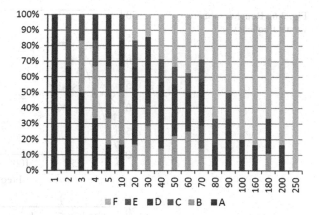

"uneven". Kansei keyword "roughness" can be felt at broad area from 4 to 200 Hz and "fine" can be felt under 100 Hz.

For the next experiment, the relationships between the frequency distribution of vibratory stimulus and smoothness/roughness are investigated. In the experiments, the input waves are used as the combination patterns (named p1 and p2) and one wave is displayed to the subject for 3 s, and after 0.2 s the other wave is displayed to the subject for 3 s, sequentially. After that, the subject evaluate smoothness/roughness of p2 compared with p1 using the evaluation form with five grades as shown in Fig. 3.5.

In the experiments, the experiment orders are random and the displayed order of p1 and p2 are also. At the experimental time, the subjects don't know what the displayed stimuli are. The amplitude of the displayed stimuli is settled at 30 μm and the frequencies of the stimulus were changed from 1 to 250 Hz. The subjects are seven male and they are 21–32 years old.

By the Scheffe's pair comparison method [2], one of semantic differential methods, the scores of each wave pattern are obtained. Analysis results of pair comparison are shown in Fig. 3.6. Figure 3.6 shows score versus stimulus frequency, and the results are separated roughly into three groups. In the figures, the higher score means that human evaluate the displayed stimulus is rougher and the lower score means that human evaluate the displayed stimulus is smoother.

The subjects of the first group feel the roughest from 50 to 100 Hz, and those of the second group feel the roughest at about 200 Hz, and that of the last group feel the roughest twice at 50 and 200 Hz. These results are caused that Meissner's

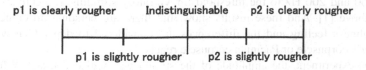

Fig. 3.5 Evaluation form of sensory test of the roughness

Fig. 3.6 Roughness score
versus frequency

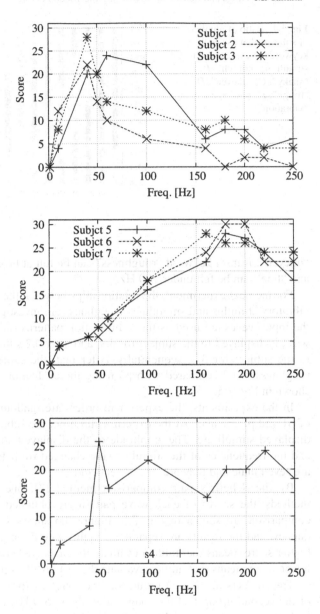

corpuscle and Pacinian corpuscle have the highest sensitivity frequency area
around 50 and 200 Hz to the threshold for vibratory stimulus, respectively mentioned above [1]. And these results show that there are personal differences for
the roughness feeling and the differences are considered to depend on whether
Meissner's corpuscle or Pacinian corpuscle reacts.

In this experiment, the amplitude of the stimuli was settled constant. When the
frequencies were settled constant and the amplitude of the stimulus were changed,

human touch feelings about roughness are investigated. It was confirmed that
the roughness feelings increase with the increase of the amplitude of the stimuli.
Moreover, we must consider the combination of amplitude and frequency, it is necessary to investigate the relations about amplitudes and frequencies in detail and to
increase the number of subjects for the investigation of the influence of the individual
variation.

3.6 Sensory Tests for Hardness

Next, the target is softness that is also one of fundamental touch feelings. The relationships between physical properties of soft objects and the tactile softness are
investigated. After this, the tactile softness means touch feeling of softness when
human touches an object. First, the relationship between the stiffness of measured
objects and the tactile softness is investigated using silicone blocks with different
Young's modulus.

To investigate the mechanism of evaluating tactile softness of human, the
relationship between the stiffness of evaluated objects and tactile softness when
human touch to the object and feel is investigated. In experiments, six kinds of
silicone block objects with different Young's modulus are prepared. The dimensions of these objects are 30 mm width, 30 mm length, and 20 mm thickness.
Young's modulus of them are 0.37, 0.82, 0.94, 1.01, 1.47, and 2.86×10^{-1} MPa.
Young's modulus of the objects are determined by reference to Young's modulus
of epidermis, dermis and hypodermis of skin [3–6].

Using these silicone objects, a sensory test of tactile softness is conducted. In
an experiment, two objects are picked out of the six objects, and those are placed
on the force sensor. Six subjects touch the objects using their forefinger alter-
nately, and compared tactile softness of two objects. The subjects are 20–32 years
old men. The force sensor can measure the contact force applied on the objects
vertically when a subject touches the object. The sensory tests are conducted in
total 15 combinations of the six objects. The results of sensory test were evaluated
using Scheffse' paired comparison method [2].

The subjects answered which object is higher young's modulus, and the correct answer rate was almost 100 % in all trial, and it found that the tactile softness
of the objects decrease with increase of Young's modulus of the objects. Young's
modulus and stiffness of the objects mean the same tendency, because thickness of
all objects are the same. It can be said that tactile softness of the objects decreases
with increase of stiffness of the objects.

The contact force was applied to the objects by the subjects with their forefinger in the sensory test and it was recorded by the force sensor as shown Fig. 3.7.
Figure 3.8 shows the scene of the sensory test, and one example of the force sensor output. The peak of the force is defined as the value of the contact force. The
contact force is almost 5–15 N. In order to investigate the relation of hardness and
contact force in detail, the difference and ratio are investigated.

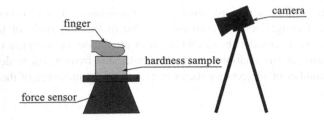

Fig. 3.7 System setup for sensory test measuring hardness

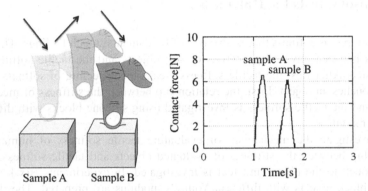

Fig. 3.8 Scene of the sensory test for hardness and one example of the contact force sensor output

Figures 3.9 and 3.10 show the differences and ratio of the contact force between the compared two objects, respectively. Sample A is harder than sample B and the young's modulus is called Ea and Eb (Ea > Eb), and Fa and Fb are the peak force of samples A and B. The horizontal axis values are difference (Ea−Eb) and ratio (Ea/Eb) of Young's modulus between compared two objects. The vertical

Fig. 3.9 Difference of Young's modulus and contact force

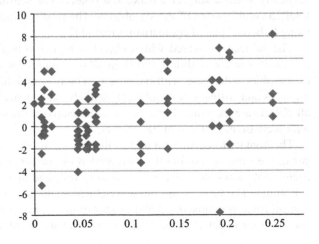

Fig. 3.10 Ratio of difference of Young's modulus and contact force

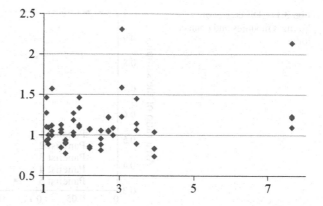

axis values are differences (Fa−Fb) and ratio (Fa/Fb) of the contact force between compared two objects.

In the figures the negative difference values and the ratio values that are smaller than one in the figures mean that the subjects touch softer object with higher contact force. The differences of the contact force vary widely, and there are some negative values and there are many ratio values that are smaller than one. As mentioned before, the correct answer ratio was about 100 %, therefore, it is said to be difficult to evaluate the tactile softness of the objects with only contact force information.

3.7 Influence of Contact Area Upon Tactile Softness Evaluation

It is confirmed that it is difficult for human to evaluate the tactile softness of the objects with only contact force information. Here, the influence of contact area upon perception of tactile softness evaluation is investigated. We consider that the perception of tactile softness is affected by the contact area information in two ways. One is a size of contact area between subjects' finger and evaluated objects and the other is variation of the contact area size in touch motion. And the latter is investigated.

At first, the relationship between contact force and contact area is investigated. Five subjects push their forefinger into three silicone blocks with ink, in such a way as to evaluate tactile softness of the blocks. The contact force is measured using a pressure sensor that is placed under the object, and the size of contact area between the forefinger and the blocks are calculated using ink blot on the blocks.

Figure 3.11 shows one of examples of the results of the experiment. The sizes of contact area are normalized using that of the softest silicone block. As the results, the size of contact area is depending on the Young's modulus and the size decrease with increase in Young's modulus of the blocks.

Fig. 3.11 Difference of
Young's modulus and contact
force

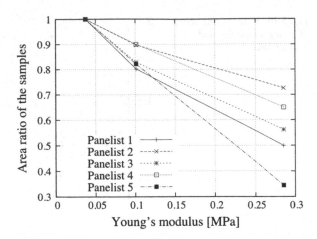

Young's modulus [MPa]

Thus, we focused on the influence of contact area between the finger and an object upon evaluation of tactile softness. Then we tried two sensory tests with four silicones under the different contact condition. Four kinds of silicone softness objects are prepared for the sensory tests. Dimensions of these objects are 30 mm width, 30 mm length, and 20 mm thickness. Young's modulus of the objects in the sensory tests are 0.37, 0.83, 1.01, and 1.47×10^{-1} MPa.

In the sensory tests, two objects are picked out of the four objects, and named object A and object B. Six subjects touch the objects using their forefinger, and compare tactile softness of the objects. The subjects are 20–32 years old men. The sensory tests are conducted in total six combinations of the four objects. Table 3.2 shows the result of the first sensory test. In the table, "s1" to "s6" mean the subjects, and the item of "A" or "B" in "Evaluation of subjects" means the object that the subject evaluated harder. Almost all subjects evaluate that the object with higher Young's modulus is harder.

Next, the subjects evaluate the tactile softness of the objects through the cylinder piston device as shown in Fig. 3.12. The cylinder piston device consists of a piston, and a stage. Shape of the contact 1 is 5 mm square. And that of the contact 2 is 10 mm square. The size of contact 1 is sufficiently-small as compared with the contact area

Table 3.2 The result of the first sensory test of tactile softness without the piston device

Young's modulus [MPa]		Evaluation of subjects					
Object A	Object B	s1	s2	s3	s4	s5	s6
0.147	0.101	A	A	A	A	A	A
0.147	0.083	A	A	A	A	A	A
0.147	0.037	A	A	A	A	A	A
0.101	0.083	A	A	A	A	A	A
0.101	0.037	A	A	A	A	A	A
0.083	0.037	B	A	A	A	A	A

Fig. 3.12 A cylinder piston device for sensory test

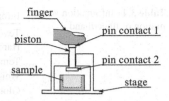

between their forefinger and the objects in the first sensory test. The piston moves vertically in accordance with the motion of a subject's forefinger. The subject pushed the objects using the device to evaluate the tactile softness of the objects. The size of contact area between their forefinger and contact 1 is constant, and the subjects evaluate tactile softness of the object without influence of contact area information. The results of the experiments are evaluated using Scheffse' paired comparison method. At the time, the subjects are asked about the difficulty of the evaluation in the first sensory test without the piston device and the second sensory test with the piston device.

Table 3.3 shows the result in the second sensory test. "Evaluation of subjects" means the object that the subject evaluated harder. Some subjects tend to evaluate that the object with lower Young's modulus is harder. The percentage of the subjects that evaluated the object with higher Young's modulus as harder in all trials is 97.2 % in the first sensory test, but that is 86.1 % in the second sensory test. It was confirmed that all subjects feel it more difficult to evaluate tactile softness of the objects in the second sensory test than that in the first sensory test. These results suggest that the contact area information is important to compare tactile softness difference between slight different objects.

3.8 Design of Sensor for Measuring Touch Sensation

For the development of the sensor system, we have focused on the three points. First point is the motion of the hand/finger. Human changes unconsciously the motion of the hands/fingers depending on the information that we want to know. Lederman et al. have studied the relationship between the information and motion [7].

Table 3.3 The result of the second sensory test of tactile softness using cylinder piston device

Young's modulus [MPa]		Evaluation of subjects					
Object A	Object B	s1	s2	s3	s4	s5	s6
0.147	0.101	A	A	A	A	A	B
0.147	0.083	A	A	A	A	A	A
0.147	0.037	A	A	A	A	A	A
0.101	0.083	A	A	B	A	A	A
0.101	0.037	A	B	A	A	A	A
0.083	0.037	B	A	A	A	A	B

Table 3.4 Information versus motion of finger/hand

Information	Motion of the hand/finger
Texture	Lateral motion
Hardness	Pressure
Temperature	Static contact
Weight	Unsupported holding
Global shape/volume	Enclosure
Global shape/exact shape	Contour following

For example, when we want to know the textures of the objects, we push softly and stroke over the surfaces of the objects. Table 3.4 shows the relationship. These indicate the design of the driving equipment of the sensor system.

Second point is to select the suitable sensor element from many sensor elements. The clarification of the mechanism of the sensory receptor in human skin and touch feelings is useful to consider which information is necessary and as a result sensor elements are chosen. For the sensor system, the chosen sensor element is mounted to driving equipment and the sensor output is obtained. The last one is the signal processing of the sensor output obtained by the sensor system. The signal processing is also derived from the clarification of the mechanism of the sensory receptor in human skin and touch feelings.

3.9 Sensor for Measuring Touch Sensation

Especially, we have focused on the Pacinian corpuscle and we have already developed some tactile sensor systems by using the feature of the Paccini. As mentioned before, Pacinian corpuscle plays important role in high-frequency vibrations that occur when we move our fingertips over structures with very fine texture.

We already tried to measure the tactile sensation of fabrics [8–10], human skin [11–16], hair conditions [17], and we succeeded. Figure 3.13 is the sensor part. PVDF film, polyvinylidene fluoride film is one of the piezoelectric materials. The film is effective to measure a force and displacement and it is used for the sensory material. In addition, PVDF film has features light weight, flexible, sensitive, and thin. The thickness of the film is 28 μm. Furthermore, the film has the feature that the response is very similar to that of Pacinian corpuscle that is one of the receptor of human skin. The lattice shape surface has a role to improve the sensitivity of the sensor as like human finger print.

Figure 3.14 is the one of the sensor system that measures the touch feelings of the fabrics. The sensor part as shown in Fig. 3.13 is attached to the tip of the artificial robotic finger and the sensor is pushed to the surface of the fabric softly by using the piezoelectric actuator. And the sensor is slid over the surface with the x-axis stage driven by a stepping motor. The obtained sensor output is transmitted to personal computer via AD card and the sensor signal is processed.

The typical sensor output is shown in Fig. 3.15.

Fig. 3.13 Sensor part

Fig. 3.14 Sensor system

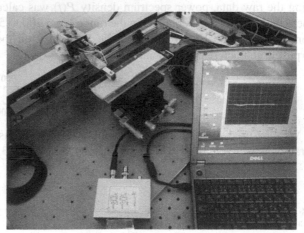

For the signal processing, we calculated two parameters. One is the evaluation of the magnitude of the amplitude of the sensor output, using this equation. This parameter is based on the feature of Pacinian corpuscle, since the output of Pacinian corpuscle is proportional to the applied force.

$$\text{Var} = \frac{1}{N-1} \sum_{i=1}^{N} (x(i) - \bar{x})^2$$

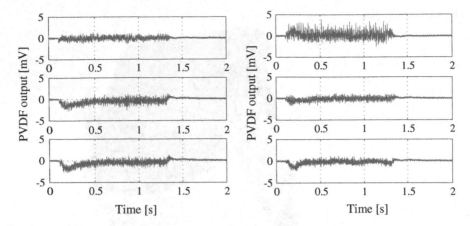

Fig. 3.15 Typical sensor output on fabrics

Here, N is the data point number of PVDF output, $x(i)$: ith PVDF output signal, and $\bar{x}$ is the average of PVDF output.

The other is distribution of the power intensity in mid-frequency range Rs. By using the raw data, power spectrum density $P(f)$ was calculated by FFT analysis. Then, the summation is obtained in the range from 100 to 500 Hz, and in the range from 100 to 2,000 Hz. We selected the frequency on the basis of the Pacinian corpuscle characteristics as mentioned before.

$$Rs = \frac{Sa}{Sb}, \; Sa = \sum_{f=100}^{500} P(f), \; Sb = \sum_{f=100}^{2,000} P(f)$$

We measured six kinds of women's underwear with this sensor system and two parameters were obtained. The plotted data is shown Fig. 3.16.

Fig. 3.16 Rs versus Var
obtained from the sensor
output

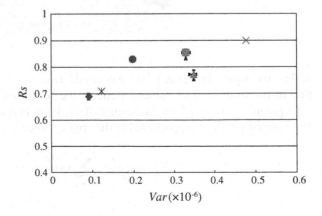

Separating from the measurement experiment, the sensory test was done. Semantic differential (SD) method with five grades scale was used, 14 key words are prepared as follows. 1: Damp–Fine, 2: Rough–Smooth, 3: Prickling–Not prickling, 4: Dry–Moist, 5: Not slimy–Slimy, 6: Not sticky–Sticky, 7: Hanging about–not hanging about, 8: Hard–Soft, 9: Not downy–Downy, 10: Not cool–Cool, 11: Warm–Not warm, 12: Not wet–Wet, 13: Bad feeling–Good feeling, 14: Uncomfortable–Comfortable. Subjects are five women, who are evaluation experts in industry, and they are 30 and 40s.

There are many kinds of analysis method, and we tried factor analysis. The elements of the first principal component are "damp", "moist", "wet", "hanging about", "sticky" and "fitting" and the elements of the second are "downy", "soft", "not prickling", "smooth", "not cool", "warm". The factor loading of the first principal component is larger than 0.9 and the loading of the second is larger than 0.5. And the cumulative contribution ratio of the first and second components is 95 %, and the evaluation of wear and/or touch feeling can be measured by the two components.

The sensor in Fig. 3.15 cannot measure the feelings of "Tight" and "Warm", therefore, as the value of the first component, the average of "damp", "moist", and "wet feeling" is calculated, and as value the second, the average of "soft", "downy", and "not prickling". The values are compared with the sensor output and we obtained the clear relations between *Rs* and downy and soft feeling whose correlation coefficient is—0.79, and Var and damp and wet feeling is—0.81.

3.10 Palpation Sensor

Palpation sensor for measuring hardness has also been developed [18]. The sensor pushes the object based on the haptics motion in Table 3.4. One example of sensor is shown in Fig. 3.17. PVDF is used also sensory material. And the amplitude

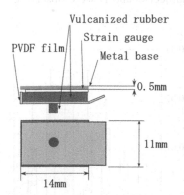
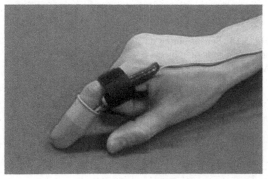

Fig. 3.17 Palpation sensor for measuring prostate

of the output is evaluated. The sensor is attached to the tip of finger and a small-sized motor is attached to the base of the finger to excite the sensor to the object. When the object is harder, the amplitude of the sensor output is larger. We tried to measure the various prostates in clinical tests, such as normal prostate, prostatic cancer, hypertrophy, with stones. The hardness of prostatic cancer is similar to that of bone, and that of prostatic hypertrophy is elastic. Concerning the prostate with stone, the hardness of the part of stone is as same as stone.

Figure 3.18 shows the result that is one example in clinical test. μ means in vertical axis the average of the amplitude of the sensor output. The prostate conditions of subject are as follows. A: Normal and healthy, B: almost normal and healthy, C: between and normal prostatic hypertrophy, D: under treatment of the prostatic cancer, E and F: prostate stones in places, G: prostate stoned.

It is seen that the outputs on subjects A and B are much smaller than the others and the output of B is slightly larger than that of A. The difference corresponds to the difference of the disease A and B.

The sensor value on subject C is much larger than that of subject A and B. It means that the state of the prostate is closer to the hypertrophy. It is seen that the sensor output on subject D, who is under treatment of prostatic cancer, is closer to that of subject C. About the condition of subject D, the palpation result of the doctor without sensor could not distinguish whether the stiffness is that of the prostatic cancer or hypertrophy. The result of the sensor output means the condition is closer to that of prostatic hypertrophy. The sensor output is effective to evaluate the disease conditions.

Subjects E and F have prostate stones in places, and the sensor output takes the maximum and much larger than the others'. Subject G has one prostate stone and it was difficult for the doctor to search for the part. Therefore, the measurements were done many times and the fluctuation of the sensor output was large. From the doctor's diagnosis, the state of prostate except the stone part is prostate hypertrophy. Therefore, the average value became the smaller than that of subject E and F. However, it is noticed the maximum value on subject G is large and the result means that it is unhealthy condition.

Further, the conditions of subjects were investigated using the ultrasound tomography. The enlarged prostate conditions of subject C and D were observed.

Fig. 3.18 Sensor output on prostate glands of subjects A–G

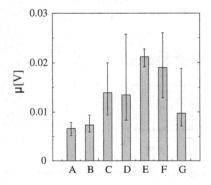

The white marks, which correspond to the prostate stones, were observed in places on the prostate glands of subject E and C. However, the white mark could not be discovered on the prostate of subject G, by ultrasound tomography.

These results showed that the output of the present sensor varies with the stiffness of prostate glands and the present sensor output has a good correlation between doctor's palpation result. Further, it is said the sensor is effective in diagnosing the condition of prostate glands.

3.11 Concluding Remarks

In this chapter, it is shown that the characteristics of human sensory and haptics of hand/finger are useful. In order to make various things that human feel comfortable and good feelings in industry, it is important to know how human feel when he/she use the made things, and the establishment of an objective evaluation method including sensor system is required. In medical welfare fields, the highly accurate palpation sensor is expected to be effective to find the part of disease in early stage and to keep the health. Further, to know the principals and mechanism of the tactile sensation leads to the development of the technology, for example, the tactile display system that can transmit someone else touch feelings and technology that can give reality using of characteristics of touch feelings. In future, the range to use the technology of the touch feelings will extend more and more.

References

1. Shepherd GM (1998) Neurobiology. Oxford University Press, Inc., New York, pp 215–221
2. Sumiko N (1970) A transform of scheffe AN˘ s method (Japanese), In: Proceeding of 11th sensory test conference, Union of Japanese Scientists and Engineers
3. Yoshikawa Y (1975) Mechanical behavior of skin and measurement way (hardness measurement of a living body and artificial judge < special story>). Meas Control 3(14):263–280
4. Kazuo Y, Hidehiko T, Shotaro O (1975) Physical characteristics of human finger. Biomechanisms 3:27–36
5. Fung YC (1993) Biomechanics: mechanical properties of living tissues, 2nd edn. Springer, New York
6. Maeno T, Kobayashi K, Yamazaki N (1997) Relationship between the structure of finger tissue and the location of tactile receptors. Trans Jpn Soc Mech Eng Series C 63(607):881–888
7. Lederman SJ, Klatzky RL (1987) Hand movements: a window into haptic object recognition. Cognitive Psychology 19:346. Copyright 1987 by Elsevier
8. Tanaka M (2002) Measurement and valuation of touch sensation: texture measurement on underclothes. Stud Appl Electromagnet Mech 12:53–58
9. Tanaka M, Numazawa Yu (2004) Rating and valuation of human haptic sensation. Int J Appl Electromagnet Mech 19:573–579
10. Tanaka Y, Tanaka M, Chonan S (2007) Development of a sensor system for collecting tactile information. Microsyst Technol 13:1005–1013
11. Tanaka M, Tanaka Y, Chonan S (2008) Measurement and evaluation of tactile sensations using a PVDF sensor. J Intell Mater Syst Struct 19:35–42

12. Tanaka M (2001) Development of tactile sensor for monitoring skin conditions. J Mater Process Technol 108:253–256
13. Tanaka M, Hiraizumi J, Leveque JL, Chonan S (2002) Haptic sensor for monitoring skin conditions. Int J Appl Electromagnet Mech 14:397–404
14. Tanaka Mami, Leveque JL, Tagami H, Kikuchi K, Chonan S (2003) The "Haptic finger"-a new device for monitoring skin condition. Skin Res Techonol 9:131–136
15. Tanaka M, Sugiura H, Leveque JL, Tagami H, Kikuchi K, Chonan S (2005) Active haptic sensation for monitoring skin conditions. J Mater Process Technol 161:199–203
16. Tanaka M, Matsumoto M, Uetake N, Kikuchi K, Leveque JL, Chonan S (2006) Development of an active tactile sensor for measuring human skin conditions. In: Sixteenth international conference on adaptive structures and technologies, pp 335–341 (DEStech Publications Inc.)
17. Okuyama T, Hariu M, Kawasoe T, Kakizawa M, Shimizu H, Tanaka M (2011) Development of tactile sensor for measuring hair touch feeling. Microsyst Technol 17:1153–1160
18. Tanaka M, Furubayashi M, Tanahashi Y, Chonan S (2000) Development of an active palpation sensor for detecting prostatic cancer and hypertrophy. Smart Mater Struct 9:878–884

Chapter 4
Eliciting, Measuring and Predicting Affect via Physiological Measures for Emotional Design

Feng Zhou and Roger Jianxin Jiao

Abstract Emotional design plays an important role in the development of products and services towards high value-added user satisfaction and performance enhancement. One critical challenge in emotional design is the measurement and prediction of affect. Most current measurement and prediction methods are affected by many biases and artifacts. For example, verbal reports only represent the sheer reflection of consciously experienced feelings. This study aimed to evaluate affect via physiological measures. Specifically, standardized affective stimuli in both visual and auditory forms were used to elicit different affective states (7 types of affect for the visual stimuli and 6 for the auditory ones). Each affective stimulus was presented for 6 s and a wide range of physiological signals were measured, including facial electromyography (EMG) (zygomatic and corrugator muscle activity), respiration rate, electroencephalography, and skin conductance response (SCR). Subjective ratings were also recorded immediately after stimulus presentation. The physiological measures show a relatively high differentiating ability in postulating affect via statistical tests and data mining-based prediction, with highest mean recognition rates of 91.47 and 71.13 % for the visual stimuli, and 91.36 and 80.66 % for the auditory stimuli, for valence- and affect-based predictions, respectively. This technological and methodological advancement offers a great potential for the development of emotional design.

Keywords Physiological measures • Affect elicitation • Measuring and prediction • Emotional design

F. Zhou (✉) · R. J. Jiao
The George W. Woodruff School of Mechanical Engineering, Georgia Institute
of Technology, Georgia 30332, USA
e-mail: fzhou35@gatech.edu

R. J. Jiao
e-mail: roger.jiao@me.gatech.edu

S. Fukuda (ed.), *Emotional Engineering vol. 2*, DOI: 10.1007/978-1-4471-4984-2_4,   41
© Springer-Verlag London 2013

4.1 Introduction

Affect is essential in the interaction process between humans and systems [1] and is now incorporated as a design parameter to improve task performance and human satisfaction [2]. Emotional design capitalizes on this perspective by conceptualizing affect-engendering products, by adapting to and by responding to human affective states to design products desirable to humans holistically. For example, products are now designed for hedonic pleasure [3], emotional responses and aspirations [4], and user experience [5], and so on, to improve customer satisfaction. In some instances, optimal performances require an appropriate arousal level [6], such as in aviation safety and repetitive work where a state of high vigilance is desirable [7]. Csikszentmihalyi [8] proposed the concept of flow that it is often necessary to maintain or prevent partic- ular affective states to have optimal performance and enjoyable user experience, such as in tutoring and training, driving, and video gaming.

Although the importance of affect as a design parameter has been well recog- nized, several fundamental questions regarding affect acquisition, measurement, and evaluation remain not well-answered: (1) How can we effectively acquire users' affective states without interference in real time? (2) How can we measure and evaluate user affect in the interaction process? (3) How can we infer customer needs for affect? Research in psychology sheds light on solving these questions. Psychologists often frame affect in two approaches: a discrete categorical per- spective and a dimensional approach. The former conceptualizes affect discretely. For example, Ekman [9] devises a list of 15 basic emotions, among which 6 are primary emotions (sadness, happiness, anger, fear, disgust, surprise). Research on the latter has found that the most commonly used dimensions are autonomic arousal (sleepy-activated) and hedonic valence (pleasure-displeasure) [10], such as Russel's circumplex model [11].

As for the first and second questions, current practice in emotional design often uses subjective methods retrospectively in terms of affective adjectives in a discrete manner, such as user interviews, focus groups or surveys [12]. A large amount of affective adjectives are first collected concerning the consumers' feelings toward a product or other affective stimuli. Then, the most relevant and appropriate terms are selected by domain experts, ranging in numbers from several dozens to several hundreds. The selected ones are further scrutinized and structured, either manually or statistically and evaluated based on n-point Likert scales (e.g., semantic differ- ential scale [13]. These methods are convenient, and amenable to statistical analy- sis. However, verbal account of feelings only captures part of affect, and is usually expressed in abstract, fuzzy, or conceptual terms [14]. Hence, work on affect elic- itation and acquisition is often based on vague assumptions and implicit inference. Moreover, they often suffer from recall and selective reporting biases if affective responses are reported retrospectively [15]. Although the affective information can be collected concurrently, the biggest problem is their interference with the task or activity when eliciting and acquiring affective responses [10]. With regard to the third question, besides the subjective methods mentioned above, another direct method is to appraise what the user is feeling from observing their non-verbal and verbal expressions and reasoning their situations [16]. However, observation is often costly, time-consuming, and sometimes disruptive for many design tasks [17].

Evidence has shown that physiological signals can differentiate basic emotions [18]. Picard et al. [19] acquired physiological data including skin conductance response (SCR), electromyography (EMG) of the jaw, BVP, and respiration of a single subject over multiple days and obtained an 81 % accuracy on recognizing eight affective states using feature-based recognition method. In addition, physiological data can be acquired in a continuous manner which is consistent with the way people perceive emotions [20] and user affective states can be evaluated in real time. Compared with subjective self-report methods, the real-time feature is particularly useful for products that seek to respond to a user's affect in ways to improve the interaction [21]. For example, Mandryk and Atkins [22] employed psychophysiological techniques to evaluate video games and proved that directly capturing and measuring autonomic nervous system activity could provide accesses to user experience. Therefore, the user-product interaction can be tailored to the individual level to optimize the user pleasure and efficiency [23]. However, it is also suggested one physiological measure alone is not adequate to give a coherent picture of what affective state is occurring within the user [24], since the relationship between psychology and physiology is not all one-to-one and in many cases, many-to-one, one-to many, or many-to-many [25]. Thus, it is necessary to be able to monitor multimodal physiological signals.

The purpose of this study is, to what extent, multiple physiological signals can be used to acquire, measure and predict user affect with standardized affective stimuli in both visual and auditory forms in real time. For one thing, evidence has shown that both types of stimuli can be used to evoke affective reactions in physiology [26, 27]. Moreover, visual and auditory information is the most frequently used stimuli presented in products and systems that might elicit affect. Also studied is the question whether there is any difference in affective responses when participants are exposed to different forms of affective stimuli. In order to do so, statistical tests and data mining methods were employed. Initial results from statistical rests showed its potential while affect prediction based on data mining methods delved into a more detailed and specific results for further proving the potential of the proposed method.

4.2 Experiment Design for Affect Elicitation

4.2.1 Participants

Forty-six college students (23 males and 23 females), including 16 Chinese, 15 Indians, and 15 Westerners (all Caucasians: five from Germany, one from France, two from the UK, and the remainder from the USA), took part in the experiment. In order to increase homogeneity within cultural groups, Chinese and Indian students were required to be born and raised in mainland China and India, respectively and to have lived in Singapore for less than 2 years; while the Westerners were exchange students at Nanyang Technological University, Singapore, for less than six months. All the participants were aged between 20 and 30 (mean = 24.4; standard deviation = 2.3). Each participant was interviewed by the experimenter to screen out mental disorder or drug and alcohol abuse. Due to technical difficulties, only data collected from 14 participants in each cultural group (i.e., $n = 42$ in the final) with gender balance were kept. Informed consent was obtained from each participant.

4.2.2 Affective Stimuli

Twenty-eight pictures were selected as the visual affective stimuli from The International Affective Picture System (IAPS) [28]. In addition, 24 sound clips were also selected as the auditory affective stimuli from The International Affective Digitized Sounds (IADS) [29]. Based on the participants' reports, the visual and auditory stimuli were categorized into seven and six groups, corresponding to different affective states respectively.

4.2.3 Apparatus

A physiological sensing system and the E-Prime software were used for data collection. The physiological sensing system included an 8-channel Biofeedback and Neurofeedback System v5.0 (Thought Technology, New York, USA) and a Myomonitor Wireless EMG System (Delsys, Boston, USA). The former was used to collect SCR and EEG. The latter were used to collect facial EMG signals (zygomatic and corrugator muscle activity). The E-Prime software was used to ensure millisecond precision data collection. Besides, auditory stimuli were presented to the participants through a pair of Altec Lansing AHP-512 headphones. The sensor placement on the participant is shown in Fig. 4.1 below:

Fig. 4.1 Sensor placement. Note that peripheral temperature and blood volume pulse were excluded from analysis in this study

①EEG
 (frontal head)

④⑥Skin conductance
 Responses (2nd & 4th fingers)

⑦Peripheral temperature
 (5th finger)

⑤Blood volume pulse (3rd finger)

②③Facial EMG
 the zygomaticus
 major (ZM②)
 corrugators
 supercilii (CS③)

⑧Respiration amplitude
 and rate (belly)

4.2.4 Physiological Measures and Features

SCR, also known as electrodermal response, is a measure of the skin's ability to conduct electricity and represents changes in the sympathetic nervous system [30]. It has been proved to be linearly related to affective arousal and independent from affective valence [26]. SCR (in μs) is measured by fastening the electrode straps around the second and the fourth fingers of the participant's left hand. Then, SCR is smoothed and log transformed [log (SCR + 1)] to normalize the distribution of the response [26]. Four temporal features of SCR were computed in Fig. 4.2: (1) The SCR amplitude, regarded as the conductivity difference among the peak point and the initiation point; (2) Latency, which is the amount of time between the stimulus and the rise of the wave; (3) The rise time, corresponding to the temporal interval between the peak and the start point; and (4) The half recovery time, i.e., the amount of time it takes for the wave to fall back to half of its amplitude.

Facial EMG measures muscle activity by detecting and amplifying the tiny electrical impulses that are generated by muscle fibres when they contract [31]. Facial EMG measures from the zygomaticus major (ZM) and corrugators supercilii (CS) are widely used for recognition of affective states [32]. It was reported that EMG varied linearly along the valence in CS and quadratically in ZM [33]. In this research, EMG electrodes are placed over ZM and CS on the left side of the face. EMG signals (in μV) are bandpass filtered from 100 to 1,000 Hz using an elliptic, short infinite impulse response (IIR) filter, amplified ($\times$ 5,000) and rectified [26]. The baselines for ZM and CS are defined as the mean activity in the one second before stimulus onset.

Then, the change scores for both ZM and CS were computed by subtracting their respective baselines from the mean responses sampled at each half-second time interval after stimulus onset for the following six seconds (i.e., 12 samples were collected). In Table 4.1, 30 features were obtained from the statistic

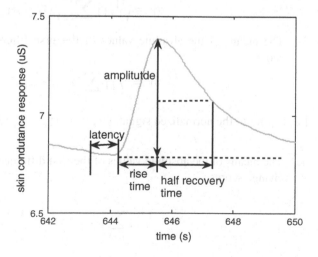

Fig. 4.2 Illustration of SCR features

Table 4.1 Features of facial EMG and respiration data

Statistic measure	ZM	CS	CZ	Integrated ZM	Integrated CS	Integrated CZ	Respiration
Mean	√	√		√	√	√	√
Standard deviation	√	√	√	√	√	√	√
MAFF	√	√	√	√	√	√	
NMAFF	√	√	√				
MASF	√	√		√	√	√	
NMASF	√	√					
Acc				√	√	√	√

Note '√' shows that the corresponding statistic measure is calculated for the physiological signal. The means of the absolute values of the first differences (MAFF), the normalized means of the absolute values of the first differences (NMAFF), the means of the absolute values of the second differences (MASF), the normalized means of the absolute values of the second differences (NMASF), the average acceleration (Acc)

features of ZM, CS, the difference score between CS and ZM change scores, i.e., CS—ZM, integrated ZM and CS change scores and CS—ZM over the presentation time. Taking ZM as an example, if Y and stand for the change scores of ZM and the i-th sample within one emotion segment $(i = 1, \ldots, 12)$, respectively, the definitions of the statistic measures are defined as follows:

1. The means of the raw change scores

$$\mu_Y = {}^1\!/_{12} \sum_{i=1}^{12} Y_i, \tag{4.1}$$

   where Y_i $(1 \le i \le 12)$ is the i-th sample of ZM or CS.

2. The standard deviations of the raw change scores

$$\sigma_Y = \sqrt{{}^1\!/_{11} \sum_{i=1}^{12} (Y_i - \mu_Y)^2}, \tag{4.2}$$

3. The means of the absolute values of the first differences of the raw change scores

$$\delta_Y = {}^1\!/_{11} \sum_{i=1}^{11} |Y_{i+1} - Y_i|, \tag{4.3}$$

Let $\tilde{Y}_i$ refer to the normalized signal, i.e., $\tilde{Y}_i = (Y_i - \mu_Y)/\sigma_Y$,

4. The means of the absolute values of the first differences of the normalized raw change scores

$$\tilde{\delta}_Y = {}^1\!/_{11} \sum_{i=1}^{11} \left| \tilde{Y}_{i+1} - \tilde{Y}_i \right| = \delta_Y/\sigma_Y, \tag{4.4}$$

5. The means of the absolute values of the second differences of the raw change
    scores

$$\gamma_Y = \frac{1}{10} \sum_{i=1}^{10} |Y_{i+2} - Y_i|, \tag{4.5}$$

6. The means of the absolute values of the second differences of the normalized
    raw change scores

$$\tilde{\gamma}_Y = \frac{1}{10} \sum_{i=1}^{10} \left| \tilde{Y}_{i+2} - \tilde{Y}_i \right| = \gamma_Y / \sigma_Y. \tag{4.6}$$

Respiration measures the rate or volume of air exchange in human lungs by its
rate or amplitude. It is observed that high arousal generally increases respiration
rate while low arousal decreases respiration [22, 34]. In this study, the respiration
sensor was sensitive to stretch and was strapped around each participant's abdo-
men. The respiration rate and relative respiration amplitude data were collected.
Empirically, the deviation respiration rate from the baseline was calculated by
subtracting the mean rate of 3 s since stimulus onset from the mean rate in 0.5 s
bins for the following 6 s. Then the mean, the standard deviation, and the average
acceleration or deceleration of the respiration measure were computed from the
sampled data as features for model construction. Note the average acceleration or
deceleration of the respiration measure is defined as follows:

$$a_R = \frac{1}{11} \sum_{i=1}^{11} (R_{i+1} - R_i) = \frac{1}{11} (R_{12} - R_1) \tag{4.7}$$

where R_i ($1 \leq i \leq 12$) is the i-th sample of respiration rate.

EEG is the recording of electrical activity along the scalp produced by the firing of neurons within the brain and reflects correlated synaptic activity caused
by post-synaptic potentials of cortical neurons [35]. It is the summation of oscil-
lations with a frequency range from 1 to 80 Hz, with amplitudes of 0.01–0.1 V
[36]. In this research, the alpha (8–12 Hz) and beta (12–30 Hz) frequencies are
extracted to reflect different affect related activities [37]. Alpha waves are typical
for an alert but relaxed mental state, visible over the parietal and occipital lobes,
while beta waves are related to an active state of mind (busy, alert, anxious thinking, and/or active concentration), most evident in the frontal cortex [36]. In this
research, the only sensing electrode is placed at the Fpz which is defined by the
international 10–20 system [38]. The two reference electrodes are located at the
left and right ear lobes. In order to minimize artefacts introduced in the EEG signals, participants are instructed not to blink during the six-second stimulus presentation as they could, and electrooculogram artefacts are corrected by an adaptive
filter based on the least mean squares regression [39] using EEGLab software
based on Matlab R2008b. The EEG data are then bandpass filtered into alpha and
beta waves using an elliptic, short IIR filter.

Alpha wave and beta wave are obtained by preprocessing the EEG data using EEGLab. Then the following energy features are calculated:

1. The power of alpha (E_α),
2. The power of beta (E_β),
3. The power of alpha and beta $(E_{\alpha+\beta})$,
4. The power of beta to alpha ratio $(E_{\beta/\alpha})$,
5. The ratio of beta power to alpha power $\left(E_\beta/E_\alpha\right)$ during the stimulus presentation.

The power spectral density (PSD) function of DelSys EMGworks analysis software (Delsys, Boston, USA) determines the PSD using the Welch method [40]. Then, the corresponding energy features are obtained within the specified (i.e., alpha, beta, and alpha + beta frequency) frequency band by integrating PSD within them. These features tell at which frequency ranges variations are strong and at which frequency ranges variations are week.

4.2.5 Procedure

First, participants were briefed that there would be different pictures (in the first place) and sound clips that might prompt different affective states and they should attend to the entire stimulus presentation for 6 s. Then, the notion of valence and arousal was explained using Self-Assessment Manikin (SAM) [41], which is a non-verbal pictorial scale (1–9, where 1 is extremely negative/calm, 5 neutral, and 9 extremely positive/activated) for the measurement of valence and arousal. Before the experiment, a practice trail with 6 pictures and 4 sound clips (these stimuli were different from those in the real experiment) was conducted by the participant. This enabled the participant to be familiar with the experiment protocols. After all the sensors were installed on the participant, a 2 min resting baseline was followed before the first stimulus was presented on a 17-inch HP desktop around 1 meter away. The experiment took place in a project room with dim lighting at 25 °C. The order of stimuli presentation was randomly but not repeatedly generated for each participant to minimize order and habituation effects. After receiving one stimulus, the participant was asked to rate the stimulus on valence and arousal using the computer-version SAM. Afterwards, each participant was asked to choose one affective adjective that best describe the affective response to the stimuli from the given terms (see Sect. 4.3). If a particular stimulus that was judged by more than 60 % of the participants to be evocative of one type of affect (e.g., sad), this stimulus was labeled as that type of affect. To avoid conflict between two consecutive affective states, there was a 10 s rest interval without any stimulus before the presentation of the next stimulus.

4.3 Affect Differentiation

4.3.1 Results for Visual Stimuli

There are 7 affective states included in this study, namely excited, amused, contented, bored, sad, fearful, and disgusted, which were grouped into three categories: positive affect (excited, amused, and contented), neutral affect (bored), and negative affect (sad, fearful, and disgusted). The corresponding pictures in the IAPS were identified as 4,645, 4,677, 4,687, 4,694 (excited), 1,340, 1,811, 2,341, 2,352.1 (amused), 2,340, 2,360, 2,550, 2,530 (contented), 7,233, 7,045, 7,010, 7,003 (bored), 2,276, 2,800, 2,900, 9,220 (sad), 6,260, 6,350, 1,120, 1,050 (fearful), and 3,080, 3,130, 3,170, 3,000 (disgusted). Each received mean votes of 78.6, 61.3, 70.8, 88.7, 86.3, 83.3, and 67.9 % from 42 participants, respectively.

In order to better differentiate among positive, neutral, and negative affect as well as different affective states, F-values of the 12 sampling data were first calculated (except subjective rating, SCR, and EEG as there was only one sample for them) in the corresponding timeline after stimulus onset and the physiological sampling data associated with maximal F-values and p-values smaller than 0.05 were selected to better distinguish among them, including facial EMG (CS and ZM) and respiration rate (Resp. Rate). Figure 4.3 shows the changes of F-values of ZM, CS, and respiration rate with their respective maximum values at the 10th, 6th, and 4th sampling data (5th, 3rd, and 5th seconds after stimulus onset, respectively) for both tests based on valence and affect with p-values smaller than 0.05.

First of all, affect was tested based on valence, i.e., positive, neutral, and negative. Table 4.2 shows the results of measures which had significant main effects, while SCR (a marginal main effect, $F (2, 291) = 2.61$, $p = 0.075$) and alpha wave ($F (2, 291) = 0.23$, $p = 0.790$) were excluded for analysis. Valence ratings had the most significant main effect, as expected; Post-hoc comparisons between any two valence categories were significantly different, $p < 0.001$ for all of them. Both positive and negative affect nearly received an equivalent amount of arousal ratings; nevertheless, they had significantly higher ratings than those of neutral affect, $p_{12} < 0.001$,

Fig. 4.3 Trends of F-values of ZM, CS, and respiration rate along sampling data for visual stimuli

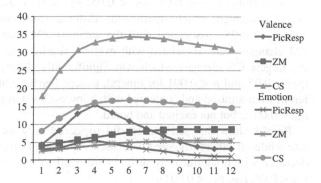

$p_{23} < 0.001$ (note p_{ij} represents the p-value of the comparison between the i-th and the j-th valence/affective states; i, j, the subscripts for each valence/affect in Tables 4.2, 4.3, 4.4, 4.5 below).

With regard to Facial EMG, both ZM and CS had main effects among three categories. ZM activity was most sensitive to positive affect and significantly distinguished between positive and neutral affect, $p_{12} < 0.01$, and between positive and negative affect, $p_{13} < 0.001$, but not between neutral and negative affect. On the contrary, the largest CS activity was elicited by negative affect and thus significantly from those by positive ($p_{13} < 0.001$) and neutral ($p_{23} < 0.001$) affect; no significant difference was found between positive and neutral affect.

Changes of respiration rate was most substantial in neutral affect which was significantly different from those in positive affect ($p_{12} < 0.001$) and those in negative affect ($p_{23} < 0.001$). With regard to beta wave, post hoc analysis showed that changes in the neutral affect and negative affect had a significant difference, $p_{23} < 0.05$.

Table 4.3 shows the results of mean values of dependent measures for 7 affective states. Power changes in alpha wave (F (6, 189) = 1.68, $p = 0.128$) and beta wave (F (6, 189) = 1.83, $p = 0.095$) failed to have significant main effects for them, and thus excluded for further analysis. As for valence ratings, consistent with the results by tests based on valence, they differed among positive, neutral, and negative affect. Further, bored and disgust significantly differed from all the other affect ($p < 0.001$ for all the pairwise comparisons); excited and contented were also significantly different ($p_{24} < 0.05$). With regard to arousal ratings, bored received lowest rating so that it differed significantly from all other states ($p_{45} < 0.05$ and $p < 0.001$ for the other comparisons). The largest ones were corresponding to excited and disgust which were significantly different from other affect except the comparison between amused and excited ($p_{13} < 0.01$, $p_{16} < 0.05$, $p_{27} < 0.05$, and $p < 0.001$ for others). Roughly consistent with arousal ratings but with a weaker main effect among different affect, the largest SCR changes were associated with excited and disgust which were significantly different from those with smallest changes, i.e., contented and bored ($p_{13} < 0.01$, $p_{14} < 0.05$, $p_{37} < 0.01$, and $p_{47} < 0.05$); however, the ones with a medium level of SCR changes, i.e., amused, sad, and fearful did not have significant differences from others.

Amused triggered the largest ZM changes and it significantly differed from all other states ($p_{12} < 0.01$, $p_{26} < 0.05$, $p_{27} < 0.01$ and $p < 0.001$ for others), except contented, which significantly differed from bored and sad ($p_{34} < 0.05$ and $p_{35} < 0.05$), associated with the smallest ZM changes. A medium level of ZM changes were associated with excited, fearful, and disgust. Disgust elicited the largest CS changes which was significantly different from all other affect ($p_{57} < 0.05$ and $p < 0.001$ for others), except fearful which was significantly different from amused and contented with a lowest level of CS changes (p_{36}, p_{28}, and $p_{46} < 0.001$), but not excited and bored.

Results show that sad and fearful triggered only increases in the respiration rate while all others triggered decreases. Among all others, bored had the biggest decrease which significantly differed from all other affect (p_{14}, p_{34}, $p_{47} < 0.01$, $p_{24} < 0.05$, p_{45}, $p_{46} < 0.001$).

Table 4.2 Mean values of dependent measures (DM) for positive, neutral, and negative affect

DM	Positive$_1$	Neutral$_2$	Negative$_3$	Main effect
Valence	7.46	5.26	2.33	$F(2, 291) = 980.78, p = 0.000$
Arousal	6.11^a	4.24	6.39^a	$F(2, 291) = 31.83, p = 0.000$
ΔZM	29.48	12.81^a	16.68^a	$F(2, 291) = 8.63, p = 0.000$
ΔCS	8.47^a	10.00^a	14.22	$F(2, 291) = 34.39, p = 0.000$
ΔResp. Rate	−0.23^a	−1.18	−0.04^a	$F(2, 291) = 15.47, p = 0.000$
ΔBeta Wave	28.68^{ab}	28.43^a	28.87^b	$F(2, 291) = 3.26, p = 0.040$

Note Letters (e.g., a, b) show the results of post hoc analysis of pairwise comparisons with Bonferroni correction within each DM and ones that share at least one letter do not significantly differ at the 0.05 level, which is also applicable to Tables 4.3, 4.4, and 4.5

Table 4.3 Mean values of dependent measures (DM) for 7 affective states

DM	Excited$_1$	Amused$_2$	Content$_3$	Bored$_4$	Sad$_5$	Fearful$_6$	Disgust$_7$	Main effect
Valence	7.38^a	7.35^{ab}	7.64^b	5.26	3.25^c	2.40^c	1.337	$F(6, 189) = 301.44$ $(p = 0.000)$
Arousal	6.82^{ab}	6.01^{ac}	5.49^c	4.24	5.44^c	6.32^c	7.40^b	$F(6, 189) = 20.74$ $(p = 0.000)$
ΔZM	18.16^{ab}	40.64^c	29.65^{bc}	12.81^a	12.73^a	20.04^{ab}	17.81^{ab}	$F(6, 189) = 5.53$ $(p = 0.000)$
ΔCS	9.11^{bc}	8.02^c	8.29^c	10.00^{abc}	11.80^{abc}	13.59^{abd}	17.27^d	$F(6, 189) = 16.70$ $(p = 0.000)$
ΔResp. Rate	−0.10^a	−0.37^a	−0.21^a	−1.18	0.07^a	0.04^a	−0.23^a	$F(6, 189) = 5.56$ $(p = 0.000)$
ΔSCR	0.131^b	0.113^{ab}	0.073^a	0.075^a	0.099^{ab}	0.116^{ab}	0.132^b	$F(6, 189) = 3.02$ $(p = 0.003)$

Table 4.4 Mean values of dependent measures (DM) for positive, neutral, and negative affect

DM	Positive$_1$	Neutral$_2$	Negative$_3$	Main effect
Valence	7.02	4.78	2.83	$F(2, 249) = 169.93, p = 0.000$
Arousal	6.54[a]	4.38	6.48[a]	$F(2, 249) = 8.01, p = 0.000$
ΔZM	28.70	13.86[a]	17.29[a]	$F(2, 249) = 6.48, p = 0.002$
ΔCS	13.98[a]	15.36[a]	25.06	$F(2, 249) = 22.71, p = 0.000$
ΔResp. Rate	−0.76[a]	−0.18	0.11[a]	$F(2, 249) = 3.174, p = 0.044$
ΔAlpha Wave	32.48[ab]	29.08[a]	34.44[b]	$F(2, 249) = 4.02, p = 0.019$
ΔBeta Wave	29.11[a]	25.72	30.64[a]	$F(2, 249) = 9.48, p = 0.000$

Table 4.5 Mean values of dependent measures (DM) for 6 affective states

DM	Happy$_1$	Excited$_2$	Bored$_3$	Sad$_4$	Disgust$_5$	Fearful$_6$	Main effect
Valence	7.07[a]	6.98[a]	4.78	3.67[b]	3.09[b]	1.73	$F(5, 246) = 222.47, p = 0.000$
Arousal	5.85[b]	7.24[a]	4.38	6.06[b]	5.92[b]	7.47[a]	$F(5, 246) = 44.10, p = 0.000$
ΔZM	33.19[a]	19.71[ab]	12.67[b]	14.31[b]	19.36[ab]	14.09[b]	$F(5, 246) = 4.23, p = 0.001$
ΔCS	10.70[a]	12.60[ab]	12.77[ab]	17.26[bc]	23.15[c]	22.61[c]	$F(5, 246) = 10.94, p = 0.000$
ΔResp. Rate	−0.46	0.22	−0.21	0.15	0.09	0.39	$F(5, 246) = 2.01, p = 0.078$
ΔAlpha Wave	33.79[ab]	31.17[a]	29.08[a]	30.81[a]	34.58[ab]	37.94[b]	$F(5, 246) = 3.83, p = 0.002$
ΔBeta Wave	28.32[ab]	29.89[acd]	25.72[b]	27.17[bd]	33.06[c]	31.69[ac]	$F(5, 246) = 8.63, p = 0.000$

4.3.2 Results for Auditory Stimuli

For the auditory stimuli, there were 6 affective states selected, namely happy, excited, bored, sad, disgusted, and fearful. The corresponding auditory stimuli in the IADS were 226, 230, 110, 221 (happy), 216, 200, 311, 815 (excited), 708, 262, 322, 723 (bored), 261, 105, 280, 293 (sad) 702, 251, 252, 255 (disgust), and 276, 277, 286, 279 (fearful) with mean votes of 71.4, 72.0, 89.9, 63.7, 75.6, and 76.8 % from 42 participants, respectively.

In a similar strategy to the visual stimuli, measures of facial EMG (CS and ZM) and respiration rate (Resp. Rate) that best differentiate among 6 states were identified by selecting the maximum F-values of the 12 sampling data. Figure 4.4 shows that, for tests based on valence, the maximum F-values for ZM, CS, and respiration rate were at 12th, 12th, and 8th sampling data (6th, 6th, and 7th seconds after stimulus onset), respectively, while for tests based on states, those maximum values were at 11th, 10th, and 3rd sampling data (5.5th, 5th, and 4.5th seconds after stimulus onset), respectively.

Table 4.4 shows that each measure had a significant main effect, except SCR (F (2, 249) = 1.07, $p > 0.05$). Valence ratings had the most significant main effect, where all comparisons of ratings from any two categories were significant, $p < 0.001$. Neutral received the lowest arousal ratings, significantly different from those of positive ($p_{12} < 0.001$) and negative ($p_{23} < 0.01$) affect while positive and negative affect received nearly equivalent ratings.

Similar to facial EMG activity for the visual stimuli, ZM activity was the most sensitive to positive affect and significantly distinguished between positive and neutral affect, $p_{12} < 0.01$, and between positive and negative affect, $p_{13} < 0.01$, but not between neutral and negative affect. On the contrary, CS activity was the most active for negative affect and significantly differed between negative and positive affect $p_{13} < 0.001$, and between neutral and negative affect $p_{23} < 0.001$.

Unlike changes of respiration rate for the visual stimuli, respiration rate increased for negative affect but decreased both for positive and neutral affect. Significant difference was only found between negative affect and positive affect, $p_{13} < 0.05$. For EEG, both power changes of alpha and beta waves had main effects and post hoc analysis showed that for alpha wave, changes in neutral affect

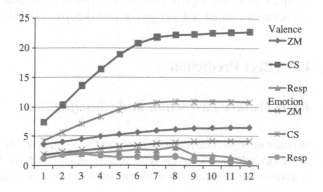

Fig. 4.4 Trends of F-values of ZM, CS, and respiration rate along sampling data for auditory stimuli

and negative affect had a significant differences, $p_{23} < 0.05$ while for beta wave, changes in neutral affect were significantly from those in positive ($p_{12} < 0.05$) and negative ($p_{23} < 0.001$) affect.

Table 4.5 shows the results of mean values of dependent measures of 6 affect for the auditory stimuli. Valence ratings significantly differed among three categories: positive (happy and excited), neutral, and negative (sad, disgust, and fearful) affect ($p < 0.001$ for all the pairwise comparisons). However, it did not distinguish between happy and excited which received the highest ratings, and between sad and disgust which received second lowest ratings after fearful. There were three levels of arousal ratings, the highest arousal ratings associated with excited and fearful, the medium arousal ratings, i.e., happy, sad, and disgust, and the lowest arousal rating, i.e., bored; comparisons between any two from different levels were significantly different from each other, $p < 0.001$ for all of them.

With regard to ZM activity, happy triggered the largest changes which significantly differed from bored, sad, and fearful (p_{13}, p_{14}, $p_{16} < 0.01$); excited and disgust triggered a medium level of ZM changes; bored and sad triggered the smallest changes in ZM; however, the latter two levels did not significantly differ with each other. CS changes seemed inversely proportional to valence ratings of affect. The largest CS changes were associated with fearful and disgust which were significantly differently from happy, excited, and bored ($p < 0.001$ for all the comparisons involved); followed by sad with medium changes in CS which only significantly differed from happy with lowest changes in CS ($p_{14} < 0.05$).

Only a marginal main effect was found for respiration rate changes of all the affect, $F(5, 246) = 2.01$, $p < 0.10$. However, both power changes in alpha and beta waves had significant main effects. Regarding alpha wave, three levels of power changes were identified. Fearful triggered the largest power changes and significantly differs from excited, bored, and sad associated with lowest changes (p_{26}, $p_{46} < 0.05$, $p_{36} < 0.01$); disgust and happy elicited medium power changes in alpha wave without significant differences with other two levels of changes. Likewise, three levels of power changes in beta wave were recognized. Disgust elicited the largest changes and was significantly different from bored, sad, and happy associated with the lowest level of changes ($p_{15} < 0.01$ and p_{35}, $p_{45} < 0.001$); the medium level of changes were associated with fearful and excited which were significantly different from bored and sad, except the comparison between excited and sad (p_{23}, $p_{46} < 0.05$, $p_{36} < 0.001$).

4.4 Affect Prediction

4.4.1 Feature Extraction and Classification

In order to predict affect, it is important to synthesize diverse measures and identify most relevant features. The results of ANOVA in Sect. 4.3 showed which features were significantly different between two or more classes. There are two popular methods with regard to feature selection and dimensionality reduction, namely principal component analysis (PCA) and linear discriminant analysis (LDA) [42]. These methods are used to find linear combination of features which best separate two or more classes of objects. However, compared with PCA, LDA explicitly makes class information to maximize ratio of between-class and within-class scatter matrix of transformed features. Therefore, LDA was first performed on a collection of features calculated in Sect. 4.2.4 for both the visual and auditory stimuli for valence- and affect-based prediction. For valence-based prediction, 3 transformed features were obtained in the LDA hyperplane while for affect-based prediction, 7 transformed features were obtained for both the visual and auditory stimuli. As an example, Fig. 4.5 shows the projection of the transformed feature set onto a two dimensional scatter plot.

Three data mining methods were then used to predict affect based on the transformed features, i.e., k-nearest neighbor (k-NN), decision rule (DR), and decision tree (DT) based on rough set theory using the rough set software system (RSES 2.2.2) (http://logic.mimuw.edu.pl/~rses/) [43]. k-NN constructs a distance measure on the basis of training samples and tested samples will make the decision based on the k ($k = 5, 10, 20$ in the test experiment) training samples that are nearest to the tested ones with respect to the calculated distance [44]. Decision rules are generated based on reducts and the predecessor of the rule takes the conjunction of certain features values or intervals and the successor takes on specific affect. These rules make it possible to predict affect with a voting strategy [45]. Decision trees are used to split feature set into fragments not larger than a predefined size as leafs which are supposed to be more uniform and easier to predict affect [43].

Fig. 4.5 2D scatter plot of the transformed feature set for visual stimuli valence-based prediction

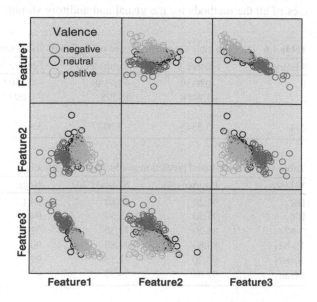

Before testing, the feature set was tabulated as a decision table, $T = (F \cup D, C)$, where F is the feature set organized in the vector form $\mathbf{f}$, D the decision set, either valence or affect. $F \cup D$ the universe of inference C, that is, $c = \{C_m^*\}_M \in C$, where M denotes the total number of the training patterns in the decision table. A specific entry of T, $C_m^* \sim \left(F_j^* \Rightarrow D_k^*\right)$, embodies the inference relationship from the features F_j^* to the corresponding affective state D_k^*. $F_j^* = \{f_{ij}^*\}_J$ is the value set of the feature vector $\mathbf{f}$, where $\mathbf{f} = \{f_i\}_I$, I the total number of features; $D_k^* = \{d_k\}_K$ (K the total number of valence or affect), the value set of D, in the test cases, was one type of particular '*valence*', or '*affect*' for the two types of predictions, respectively. The former assumed one of the symbolic values from $D^{v*} = \{positive, neutral, negative\}$ for both the visual and auditory stimuli, while the latter assumed one of the symbolic values from $D^{visual*} = \{excited, amused, content, bored, sad, fearful, disgust\}$ for the visual stimuli and $D^{auditory*} = \{excited, happy, bored, sad, fearful, disgust\}$ for the auditory stimuli. As an example, the decision table for valence-based prediction of the visual stimuli is shown in Table 4.6. To further get better prediction results, the features were discretized using cuts that were produced with the global strategy based on the maximal discernibility heuristics [45].

4.4.2 Prediction Results

In the 4 predictions, there were 294 training entries (42 participants by 7 affect) in the decision table for the visual stimuli and 252 training entries in the decision table for the auditory stimuli (42 participants by 6 affect). A tenfold cross validation was adopted for all the methods. Tables 4.7, 4.8, 4.9, 4.10 show the prediction accuracies of all the methods for the visual and auditory stimuli. As for the valence-based

Table 4.6 Decision table for valence-based prediction of visual stimuli

# Entry	Feature 1	Feature 2	Feature 3	Decision (valence)
1	−0.0339	−1.7348	−0.7502	Positive
2	−1.6468	0.4395	−0.6939	Positive
...	...	...	...	...
294	−3.5345	−3.0927	3.8757	Negative

Table 4.7 Valence-based prediction results for visual stimuli

Valence	Positive (%)	Neutral (%)	Negative (%)	Mean (%)
DR	92.20	65.80	94.10	84.03
DT	97.50	82.50	94.40	91.47
5-NN	88.20	79.30	92.50	86.67
10-NN	89.80	74.80	90.10	84.90
20-NN	90.70	71.50	92.50	84.90
Mean	91.68	74.78	92.72	86.39

Table 4.8 Valence-based prediction results for auditory stimuli

Valence	Positive (%)	Neutral (%)	Negative (%)	Mean (%)
DR	96.00	80.20	97.90	91.36
DT	91.30	75.80	96.10	87.73
5-NN	91.90	86.70	92.21	90.27
10-NN	88.90	87.20	93.30	89.80
20-NN	87.40	83.00	93.60	88.00
Mean	91.10	82.58	94.62	89.43

Table 4.9 Affect-based prediction results for the visual stimuli

Valence	Excited (%)	Amused (%)	Content (%)	Bored (%)	Sad (%)	Fearful (%)	Disgust (%)	Mean (%)
DR	90.02	55.70	36.00	84.40	79.80	60.50	81.90	69.76
DT	70.00	69.20	40.00	73.33	90.00	47.50	50.08	62.88
5-NN	67.30	55.40	56.10	83.70	60.70	57.00	79.20	65.63
10-NN	72.50	52.00	60.80	89.70	71.70	64.70	86.50	71.13
20-NN	79.50	44.50	53.30	94.60	61.30	64.20	83.60	68.71
Mean	75.86	55.36	49.24	85.15	72.70	58.78	76.26	67.62

Table 4.10 Affect-based prediction results for auditory stimuli

Valence	Happy (%)	Excited (%)	Bored (%)	Fearful (%)	Sad (%)	Disgust (%)	Mean (%)
DR	54.10	95.00	69.50	85.70	78.00	77.00	76.55
DT	63.30	100.00	72.50	84.00	86.80	73.30	79.98
5-NN	68.00	89.70	65.00	79.00	93.00	86.00	80.12
10-NN	68.60	78.30	79.70	83.33	86.80	87.20	80.66
20-NN	58.50	90.00	66.20	81.50	89.80	84.50	78.42
Mean	62.50	90.60	70.58	82.71	86.88	81.60	79.14

prediction, in Tables 4.7 and 4.8, it was found that the mean prediction rates ranged from 84.03 % by DR to 91.07 % by DT and from 87.73 % by DT to 91.36 % by DT for the visual and auditory stimuli, respectively, while the results by k-NN methods were in between; As for valence, the prediction rates across different methods of neutral were lowest whilst those of positive and negative were nearly equivalent for two types of stimuli. However, the overall results for the auditory stimuli were slightly better than those for the visual stimuli. As for affect-based prediction in Tables 4.9 and 4.10, the mean prediction rates by 10-NN obtained the best results, 71.13 and 80.66 % for the visual and auditory stimuli, respectively; the lowest prediction results were obtained by DT (62.88 %) and DR (76.55 %) for the visual and auditory stimuli, respectively. With regard to the affect, for the visual stimuli, mean prediction rates across different methods of amused, contented, and fearful were among the lowest, 49.24, 55.36, and 58.78 %, respectively, followed by sad, excited, disgust, and bored which gained the best mean prediction rate of 85.15 %. For mean prediction rates across all the methods for auditory stimuli, the lowest ones were those for happy (62.50 %), bored (70.58 %), followed by disgust, fearful, sad, and excited which was the best prediction rate at 90.06 %. It was observed that prediction errors were mainly made in the same valence category for the affect-based prediction. For example, it was hard to distinguish between amused and contented in Table 4.9 as well as sad and fearful in Tables 4.9 and 4.10. Similarly, the overall results of the auditory stimuli were better than those of the visual stimuli.

4.5 Disscussions

The current results showed that by measuring the facial EMG (ZM and CS) activity, respiration rate, SCR, and EEG (alpha and beta waves) power, as well as subjective ratings on valence and arousal, it was likely to distinguish among different valence and affective states reasonably well. For the visual stimuli, it was found that physiological measures, including ZM and CS, respiration rate, and beta wave had significant main effects when differentiating among positive, neutral, and negative affect while SCR only had a marginal main effect; A slightly difference was that SCR had a significant main effect whereas power changes in beta wave had a marginal main effect when testing among different affective states. As for the auditory stimuli, all the physiological measures had significant main effects for two types of tests, except SCR and respiration rate for the test of affective states with a marginal main effect. Further, consistent with the results of statistical tests, results of valence-based prediction were better than those of affect-based results which only obtained reasonable results. At a fundamental level, the physiological system reflect corresponding categories of affective reactions: positive affect and approach behavior, and negative affect and avoidance [46] while much smaller differences exist for affect within the same valence; Although no one measure can absolutely distinguish all the provided affect and some responses are typically activated in specific contexts whereas others remain inactive, it is possible that the combination of the multiple measures provides a big picture of affective patterns. On the one hand, based on the obtained results, it is plausible that major appraisal outcomes and major interaction patterns, although not invariably linked, maintain a non-arbitrary coherence [47].

It was demonstrated that both the visual and auditory stimuli were able to elicit affect and had relatively similar results in terms of statistical tests while better results were gained for data mining-based affect prediction for the auditory stimuli. This suggests that physiological systems are primarily sensitive to affective activation rather than to the specific mode of presentation [48]. Comparing Fig. 4.3 with Fig. 4.4, it was observed that the maximum F-values of facial EMG and respiration rate appeared at a later time (close to the stimulus offset) for the auditory stimuli. This indicates that these physiological activities continue to be active during the auditory stimulus presentation time whereas pictures do not which led to better results for the auditory stimuli. This might be because the auditory stimuli change dynamically and new affective information is serially added [48]. Another good reason is that, compared with the 7 states in affect-based prediction for the visual stimuli, there were 6 affective states for the auditory stimuli.

From a technological perspective, the physiological sensor and computing system provides one means of monitoring, quantifying and representing users' affective states in real time. Based on the mechanisms presumed to underlie the elicitation and differentiation of affect, affect is a multi-componential phenomenon, including efferent physiological, behavioral, and subjective manifestations, unfolding over time with possibly highly organized response profiles [14], multiple measures can be used to predict users' affective states more objectively and with a high ecological validity. Using wireless physiological sensors, during the human-product interaction, users' affect transitions can be uncovered without interfering with users or retrospective bias recalls that self-reports often has so that the mapping between affective needs and design elements in product can be constructed. However, self-reports can be used afterwards as further validation. Therefore, adaptive systems based on physiological computing, on the one hand, have potential to design technology that sustains user engagement and maximizes the positive affect experienced by the user, and on the other hand, minimize negative affect by offering assistance if the user is unable to perform the task [14]. And physiological monitoring systems integrated into the fabric of the wearer that continuously acquire multiple physiological data in an unobtrusive way are developed and continue to be improved, such as 'Smart Shirt' [49]. Therefore, the future of physiological measurement in emotional design is promising.

However, emotional design still remains at an early stage. Various problems yet need to be tackled. As mentioned early there is no consensus among psychologists on the framework of affect (discrete versus dimensional). Hence, it seems that models proposed here on the relationships between physiology and affective states cannot be applicable to others due to different users (influenced by age, culture, and other factors), contexts, and tasks involved. And the affective states evoked during different human-product interactions can be quite different. Therefore, it is critical that the system designer accurately assess the possible affective states, its effects on the user, and its task performance in a particular human-product interaction scenario [46]. Further, there are still some inconsistencies remain among findings reported in literature [50]. There are also differences between individuals in their degree of physiological responses and even at different occasions within the same individual. To some extent, normalization and standardization techniques can mitigate the differences among individuals. In other cases, there is specially designed product that is customized for only one particular user where the reliability of this technology can be enhanced. Another difficulty is that it is also argued that tighter experimental control with strong affective stimuli may well have produced statistically significant difference whereas it seems unlikely to have similar results under the circumstances of loosely controlled situations with subtle affective stimuli [51]. However, it is worth emphasizing that technological advances may permit one to venture into these subtle physiological changes as affective responses.

4.6 Conclusions

Emotional design focuses on users' affective needs as well as task performance which calls for which affective considerations must be measured and predicted, and how and when to respond accordingly for the system. In this study, multiple affective responses were measured using facial EMG (zygomatic and corrugator muscle activity), respiration rate, electroencephalography, and SCR as well as subjective rating when participants were exposed to a wide range of affective stimuli both visually and auditorily. Results showed that physiological signals are capable of measuring and predicting affective states in real time. In this sense, measurements based on physiological signals have the potential to associate users' affective states with physiological signals in a particular time and further with system components or stimulus events that cause these affective states. It thus extends the interaction medium between users and systems. Although the study has been conducted in lab with controlled experiments, the proposed concept has meaningful implications in design systems affectively to improve user satisfaction and task performance.

References

1. Picard RW (1997) Affective Computing. The MIT Press, Cambridge
2. Helander MG, Tham MP (2003) Hedonomics-affective human factors design. Ergonomics 46(13/14):1269–1272.
3. Helander MG, Khalid HM (2006) Affective and pleasurable design. In: Salvendy G (ed) Handbook of human factors and ergonomics, 3rd edn. Wiley, New York
4. Jiao J, Xu Q, Du J, Zhang Y, Helander MG, Khalid HM et al (2007) Analytical emotional design with ambient intelligence for mass customization and personalization. Int J Flex Manuf Syst 19:570–595
5. Desmet PMA, Hekkert P (2007) Framework of product experience. Int J Des 1(1):57–66
6. Yerkes RM, Dodson JD (1908) The relation of strength of stimulus to rapidity of habit-formation. J Comp Neurol Psychol 18:459–482
7. Wickens CD, Hollands JG (1999) Engineering psychology and human performance, 3rd edn. Prentice Hall, New Jersey
8. Csikszentmihalyi M (1990) Flow: the psychology of optimal experience. Harper and Row, New York
9. Ekman P (1999) Basic Emotions. In: Dalgleish T, Power M (eds) Handbook of cognition and emotion. Wiley, Sussex
10. Detenber BH (2001) Measuring emotional responses in human factors research: some theoretical and practical considerations. Paper presented at the international conference on affective human factors design, Singapore
11. Russell JA (2003) Core affect and the psychological construction of emotion. Psychol Rev 110(1):145–172
12. Nagamachi M (1995) Kansei engineering: a new ergonomic consumer-oriented technology for product development. Int J Ind Ergon 15(1):3–11
13. Osgood CE, Suci GJ, Tannenbaum PH (1957) The measurement of meaning. University of Illinois Press, Urbana
14. Grandjean D, Sander D, Scherer KR (2008) Conscious emotional experience emerges as a function of multilevel, appraisal-driven response synchronization. Conscious Cogn 17(2):484–495

15. Stone AA, Shiffman S (1994) Ecological momentary assessment (EMA) in behavioral medicine. Ann Behav Med 16(3):199–202
16. Picard RW, Klein J (2002) Computers that recognise and respond to user emotion: theoretical and practical implications. Interact Comput 14:141–169
17. Crabtree A, Rodden T, Hemmings T, Benford S (2003) Tools for studying behavior and technology in natural settings paper presented at the UbiComp 2003
18. Ekman P, Levenson RW, Friesen WV (1983) Autonomic nervous system activity distinguishes among emotions. Sci New Ser 221(4616):1208–1210
19. Picard RW, Vyzas E, Healey J (2001) Toward machine emotional intelligence: analysis of affective physiological state. IEEE Trans Pattern Anal Mach Intell 23(10):1175–1191
20. Schiano DJ, Ehrlich SM, Sheridan K (2004) Categorical imperative NOT: facial affect is perceived continuously. Paper presented at the conference on human factors in computing systems, Vienna, Austria
21. Bailenson JN, Pontikakis ED, Mauss IB, Gross JJ, Jabon ME, Hutcherson CAC et al (2008) Real-time classification of evoked emotions using facial feature tracking and physiological responses. Int J Hum Comput Stud 66(5):303–317
22. Mandryk R, Atkins M (2007) A fuzzy physiological approach for continuously modeling emotion during interaction with play technologies. Int J Hum Comput Stud 65(4):329–347
23. Fairclough SH (2009) Fundamentals of physiological computing. Interact Comput 21(1–2):133–145
24. Scheirer J, Fernandez R, Klein J, Picard RW (2002) Frustrating the user on purpose: a step toward building an affective computer. Interact Comput 14(2):93–118
25. Cacioppo JT, Tassinary LG (1990) Inferring psychological significance from physiological signals. Am Psychol 45(1):16–28
26. Lang PJ, Greenwald MK, Bradley MM, Hamm AO (1993) Looking at pictures: affective, facial, visceral, and behavioral reactions. Psychophysiology 1993(30):3
27. Partala T, Surakka V (2003) Pupil size as an indication of affective processing. Int J Hum Comput Stud 59(1–2):185–198
28. Lang PJ, Bradley MM, Cuthbert BN (2008) International affective picture system (IAPS): affective ratings of pictures and instruction manual. Technical report number A-8, University of Floridao, Gainesville
29. Bradley MM, Lang PJ (2007) The international affective digitized sounds (2nd Edition; IADS-2): affective ratings of sounds and instruction manual. Technical report number B-3, University of Floridao, Gainesville
30. Bernstein AS (1969) The orienting response and direction of stimulus change. Psychon Sci 12:127–128
31. Shelley K, Shelley S (2001) Pulse oximeter waveform: photoelectric plethysmography. In: Lake C, Hines R, Blitt C (eds) Clinical Monitoring. Saunders Company, Philadeiphia, pp 420–428
32. Laparra-Hernández J, Belda-Loisa JM, Medinaa E, Camposa N, Povedaa R (2008) EMG and GSR signals for evaluating user's perception of different types of ceramic flooring. Int J Ind Ergon 39(2):326–332
33. Larsen JT, Norris CJ, Cacioppo JT (2003) Effects of positive and negative affect on electromyographic activity over zygomaticus major and corrugator supercilii. Psychophysiology 40:776–785
34. Stern RM, Ray WJ, Quigley KS (2001) Psychophysiological recording, 2nd edn. Oxford University Press, New York
35. Creutzfeldt OD, Watanabe S, Lux HD (1966) Relations between EEG phenomena and potentials of single cortical cells. I. Evoked responses after thalamic and epicortical stimulation. Electroencephalogr Clin Neurophysiol 20:1–18
36. Kandel ER, Schwartz JH, Jessell TM (2000) Principles of neural science, 4th edn. McGraw-Hill, New York
37. Niemic CP (2002) Studies of emotion: a theoretical and empirical review of psychophysiological studies of emotion. J Undergrad Res 1:15–18

38. Niedermeyer E, Lopes da Silva F (2004) Electroencephalography: basic principles, clinical applications, and related fields, 5th edn. Lippincott Williams & Wilkins, Philadelphia
39. Haykin S (1996) Adaptive filter theory, 3rd edn. Prentice-Hall, Inc., Upper Saddle River, NJ, USA
40. Welch PD (1967) The use of fast fourier transform for the estimation of power spectra: method based on time averaging over short, modified periodograms. IEEE Trans. Audio Electro Acoustics AU-15:70–73
41. Lang PJ (1980) Behavioral treatment and bio-behavioral assessment: computer applications. In: Sidowski JB, Johnson JH, Williams TA (eds) Technology in mental health care delivery systems. Ablex, Norwood, pp 119–137
42. Martinez AM, Kak AC (2001) PCA versus LDA. IEEE Trans Pattern Anal Mach Intell 23(2):228–233
43. Bazan J, Szczuka M (2000) RSES and RSESlib—a collection of tools for rough set computations. Lecture notes in artificial intelligence, vol 3066. Springer, Heidelberg, pp 592-601
44. Gora G, Wojna A (2002) RIONA: a new classification system combining rule induction and instance-based learning. Fundamenta Informaticae 51(4):369–390
45. Pawlak Z (1991) Rough sets: theoretical aspects of reasoning about data, 1st edn. Springer, London
46. Hudlicka E (2003) To feel or not to feel: the role of affect in human-computer interaction. Int J Hum Comput Stud 59:1–32
47. Davison RJ, Scherer KR, Goldsmith HH (eds) (2003) Handbook of affective sciences. Oxford University Press, New York
48. Bradley MM, Lang PJ (2000) Affective reactions to acoustic stimuli. Psychophysiology 37:204–215
49. Pandian PS, Mohanavelu K, Safeer KP, Kotresh TM, Shakunthala DT, Gopal P et al (2008) Smart vest: wearable multi-parameter remote physiological monitoring system. Med Eng Phys 30(4):466–477
50. Cacioppo J, Bernston G, Larson J, Poehlmann K, Ito T (2000) The psychophysiology of emotion. In: Lewis M, Haviland-Jones J (eds) Handbook of emotions, 2nd edn. Guilford Press, New York, pp 173–191
51. Ward RD, Marsden PH (2004) Affective computing: problems, reactions and intentions. Interact Comput 16:707–713

Chapter 5
Sensory Stimulation of Designers

Céline Mougenot and Katsumi Watanabe

Abstract This chapter examines the role of designers' own experience and perception in the process of designing new products, based on an experimental approach with designers. So far, most design studies have investigated the role of visual stimuli and visual modality in the design process. Designers being humans with senses, we claim that other sensory modalities might affect the design process and outcomes. We propose an approach to study the practice of designing where both creativity and designers' sensory impressions are investigated jointly.

5.1 Role of Sensory Stimulation in the Design Process

To remain competitive, companies worldwide strive to design and manufacture distinguishable products that attract consumers. In this chapter, we focus two major factors of product attractiveness: creativity and emotional value. Recently, there has been a growing number of research investigations aimed at understanding the process of both creativity and creation of emotional value and at enhancing them. Both topics are usually studied independently, while we suggest that there is a link between design creativity and emotional value. This chapter reviews a selection of studies on design creativity as well as studies on creation of emotional value and tries to build a theoretical framework for experimental studies on the relationship between design creativity and creation of emotional value.

5.1.1 Visual Stimulation Supports Design Creativity

One of the most accepted models of creativity is the model by Amabile [1]; this model suggests that creativity is made of three components:

C. Mougenot (✉)
Department of Mechanical Science and Engineering, Tokyo Institute
of Technology, 2-12-1 Ookayama, Meguro-ku, Tokyo 152-8550, Japan
e-mail: mougenot@mech.titech.ac.jp

K. Watanabe
Research Center of Advanced Science and Technology, The University
of Tokyo, 4-6-1 Komaba, Meguro-ku, Tokyo 153-8904, Japan

S. Fukuda (ed.), *Emotional Engineering vol. 2*, DOI: 10.1007/978-1-4471-4984-2_5,
© Springer-Verlag London 2013

- Expertise: knowledge-technical, procedural and intellectual expertise
- Motivation: intrinsic and extrinsic motivation
- Creative thinking skills: how flexibly and imaginatively people approach problems (Fig. 5.1)

The «creative thinking skills» can be affected by various parameters including the type of design problem to be solved and the type of external information the designers use along the design process.

Based on this description, several studies have explored how far visual or textual information impact the level of creativity of the design outcomes [2, 4, 5], but very few investigations have dealt with the emotional processes that underlie the effects of external information.

An investigation described the mental process of analogical reasoning of architects [8]. Another has shown how designers transform words into mental images then finally into product images, with several moves of abstraction levels [11]. Sketching is an important part of design practice and design education, and there has been a large amount of research investigations on the sketching activity by designers. As [14] defined it, design can be seen as a reflective conversation based on a generation-visualization loop, made possible by the production of handmade drawings. Based on observations of the usual practice of sketching, it was confirmed by later studies that sketching would enable designers to visualize and interpret their ideas and the visualization of their own sketches would give a new twist to their idea flow [15].

In the design process, visual information is a major support for analogy-making [13]. Two famous examples are "Juicy Salif" lemon squeezer would have been inspired by sci-fi comics by Philippe Starck as a young boy [9] and "The Dancing Building" in Prague, designed by Frank O. Gehry from a movie dancing scene with Fred Astaire and Ginger Rogers [6] (Fig. 5.2).

Aside these famous examples of visual inspiration, it has been frequently observed by researchers that designers intensively browse images from magazines or Web sites and used collection of precedents. To support this common practice

Fig. 5.1 Three components of creativity [1]

Fig. 5.2 "Dancing Building" in Prague by Frank O. Gehry based on [6]

among designers, design-dedicated softwares or interfaces have been developed to retrieve images, such as Product World [12], CRAI [6], Cabinet [7], TRENDS [10].

A study with four architectural designers who had to describe verbally their mental path [8] observed that 5.8 analogies/hour were made. Some studies have demonstrated that images have a positive impact on design creativity. For instance, [3] showed that the use of visual stimuli helped designers to produce more outputs; visual stimuli lead to the production of more ideas. Also, [5] compared the level of creativity of designers' outputs, in two different working environments: with or without visual stimuli surrounding the designers. This study showed that visual stimuli not only help designers to produce more ideas but also help designer to produce outputs with a higher level of creativity.

5.1.2 Designers are Looking for Sensory Stimulation

As we could observe in a previous investigation with professional designers at FIAT and Bertone companies [10], designers claim to look for creative inspiration in various domains and various life experiences. We asked designers to list their sources of inspiration and to report activities they do in order to be creative. Their replies covered a variety of sources, such as cinema, food, music (Fig. 5.3).

More precisely, designers mentioned in a series of interviews that "looking for inspiration" during a design project means trying to find new ideas but also experiencing emotions. In fact, in our experiments where designers had to select images they found "inspirational," we could observe that designers were not only looking at images that would help them for analogical reasoning, but also images that would create an emotional impact.

When looking at pictures of which designers had to evaluate the utility value in a given design task, all designers commented the pictures in terms of emotional impact they experienced (Fig. 5.4).

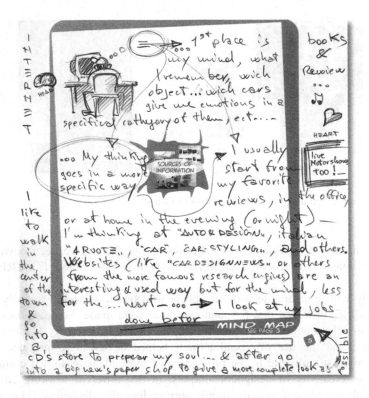

Fig. 5.3 Extract of interviews with professional designers at FIAT [10]

Fig. 5.4 Experiment: a designer is annotating "useful" images [10]

They selected these pictures because they could feel an impression of "freshness," "coolness," "simplicity" and so on. In this case, the visual stimuli caught designers' eyes because they helped to experience a feeling or an emotion (Fig. 5.5).

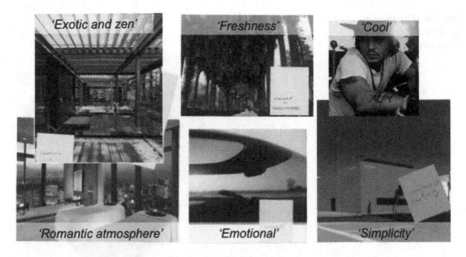

Fig. 5.5 Selected images are annotated with explanations about their "usefulness"

5.1.3 Emotion, Fourth Component of Creativity?

Much effort was put on understanding how stimuli could support design creativity and analogical reasoning. However, how to support the creation of emotional value was not much investigated (Fig. 5.6).

As we saw earlier, the visual modality has been largely studied because designers are known to intensively use visual sources of inspiration (magazines, pictures, mood boards...). Other sensory modalities have remained outside the scope of design investigations so far. Yet, designers are human beings with their own impressions and emotions which are affected by all sensory modalities, and we believe that designers' experience, including emotional, influences the whole design process and the quality of the design outputs. Thus, in our investigations, the focus is set on the designers' themselves, and our goal is to understand how external information affects the designers' emotional processes and level of creativity.

Our research hypotheses are the following:
H1   Sensory stimulation boosts designers' creativity
H2   Stimulation of memory recall boosts designers' creativity

5.2 Comparing the Effect of Sensory Modalities in a Creative Task

The study of design practice should not concentrate on the sole visual modality. Sound for instance is a powerful emotion trigger. This experiment tries to answer the following question: Can design creativity be enhanced by auditory stimulation?

Fig. 5.6 Emotions as
the fourth component of
creativity?

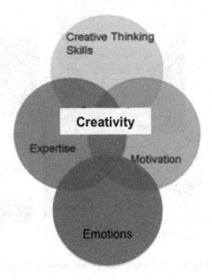

We expect that auditory stimuli will support design creativity as much as visual stimuli do but also we expect that the type of creativity supported by auditory stimuli will be different from design creativity stimulated by visual stimuli.

5.2.1 Experimental Approach

Twenty-five participants had to sketch a new concept of chair, either after having looked at a picture (n = 13) or having listened to a sound (n = 12), both representing the same situation or product (e.g. picture of fireworks in the visual condition and sound of fireworks in the auditory condition). For each of these two sensory modalities, four different types of stimuli were given: crying baby, waterfall, hairdryer and fireworks.

The novelty of each concept was assessed by two external judges (experienced designers). The assessment was based on a guideline we designed with the idea that design creativity should be reflected in artifacts with high emotional value and novel modes of usage and user interaction (Table 5.1).

As an example, a chair concept that had a novel shape but no originality in its usage received a "1." The results show that auditory stimuli, as compared to visual stimuli, tended to support the production of more original design concepts.

Table 5.1 Guideline for assessing the originality of the concepts sketched by the participants

Low-level features (look, shape, color, texture)	High-level features (use, interaction, experience)	
Not novel	Not novel	→0
Novel	Not novel	→1
Novel/not novel	Slightly novel	→2
Novel/not novel	Highly novel	→3

5.2.2 Results and Discussion

The ratings given to the sketches of each group in the four cases and throughout the whole experiment were examined. The average scores for the group with visual stimuli vary between 0.77 and 1.08, with an overall average of 0.88. Given their low scores, the sketches of this group seem to be original in terms of low-level features only, like shape, pattern or texture. The average scores for the group with auditory stimuli vary between 0.83 and 1.89, with an overall average of 1.42. This group generated more novel concepts, with a tendency toward original high-level features, like new modes of using the product or interacting with it. A two-tailed t test was performed. In all cases, the inter-judge agreement (Cronbach's alpha) was higher than 0.7, which we consider acceptable in this context (Fig. 5.7).

In the case of stimuli #1 and #2, there was no statistically significant difference between the groups. As of stimuli #3 and #4, and also throughout the whole experiment, the sketches produced with auditory stimuli received higher scores than the sketches produced with visual stimuli and this difference between groups was statistically significant (Table 5.2).

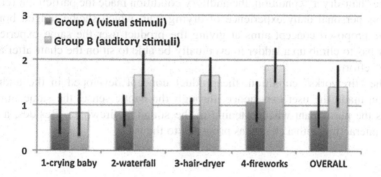

Fig. 5.7 Average originality scores

Table 5.2 Comparison of group A (visual condition) and group B (auditory condition)	Stimulus	Inter-judge agreement (Cronbach's alpha)	Group A/Group B comparison (two-tailed t test)
	1. Crying baby	0.743	$t(19) = 0.101$; NS
	2. Waterfall	0.787	$t(23) = 0.92$; NS
	3. Hair-dryer	0.8	$t(21) = 2.74$; $p = 0.012$
	4. Fireworks	0.866	$t(23) = 2.36$; $p = 0.027$
	overall		$t(91) = 2.93$; $p = 0.004$

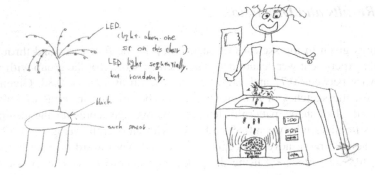

Fig. 5.8 Examples of sketches produced by participants (stimulus: "fireworks"). *Left* visual condition/*right* auditory condition

The examples in Fig. 5.8 and illustrate how the participants who received auditory stimuli generally tended to imagine new stories based on their personal experiences and to create products that embodied these stories.

In the "hair-dryer" condition, the auditory condition made the participant remind of her/his personal daily experience of drying hair, which she/he judged boring; thus, the proposed concept aims at giving the product user the same experience: the user has to climb up a ladder to eventually be able to sit on the chair after a big "boring" effort.

In the "fireworks" condition, the product concept developed in the auditory condition made the user experience, through the haptic sense, the same impressions as the participant when listening to the sound of fireworks. Besides, a new way of interacting with a chair was proposed to the user.

5.3 Discussion

The experiments showed that with auditory stimuli, the participants tended to create original experiences and new ways of interacting with the designed product; they also tended to address a wider range of senses and to create products that arouse deeper impressions and emotions among users. The analysis of the sketch annotations shows that the most original concepts were produced after a recall of personal experiences and memories which were elicited in higher number in the auditory condition. Sounds have the potential to help designing products with unique affective properties because they may afford more freedom for sketching, unlike pictures, which might tend to prime designers toward irrelevant low-level image properties like shape or color. Our experimental results support the idea that auditory stimulation boosts designers' creativity. Thus, we think it would be useful for designers to use sounds, not only images, as creative stimulation in the design workplace.

In our chapter, we proposed a research approach where both creativity and designers' experience are taken into account in the study of design protocols. Through an experimental approach reproducing usual creative process of designers, we could observe that designers' creativity was enhanced by an auditory stimulation, as compared to a visual one.

Overall, our study helped to get a better understanding of the role of designers' impressions in the design process and shows that it is necessary to stimulate designers' own affect in order to enhance design creativity. These results lead to practical recommendations for design practice, like using a large variety of stimuli from various types, visual stimuli but also auditory stimuli.

References

1. Amabile TM (1983) The social psychology of creativity. Springer, New York
2. Casakin H, Goldschmidt G (1999) Expertise and the use of visual analogy: implications for design education. Des Stud 20(2):153–175
3. Christiaans H (1992) Creativity in design: the role of domain knowledge in designing. Ph.D thesis (industrial design). TU Delft, The Netherlands
4. Cross N, Christiaans H and Dorst K (1996). Introduction: the delft protocols workshop in analysing design activity. In: Cross N, Christiaans H, Dorst K (eds.). Wiley, Chichester
5. Goldschmidt G, Smolkov M (2006) Variances in the impact of visual stimuli on design problem solving performance. Des Stud 27(5):549–569
6. Kacher S, Bignon JC, Halin G, Humbert P (2005) A method for constructing a reference image database to assist with design process. Application to the wooden architecture domain. Int J Architectural Comput 3(2):227–243
7. Keller AI (2005) For inspiration only. Ph.D thesis (industrial design). TU Delft, The Netherlands, p 175
8. Leclercq P, Heylighen A (2002) 5, 8 analogies per hour: a designer's view on analogical reasoning. AID'02 Artificial Intelligence in Design 2002, Cambridge, UK
9. Lloyd P, Snelders D (2003) What was Philippe Starck thinking of? Des Stud 24(3):237–253
10. Mougenot C, Bouchard C, Aoussat A, Westerman SJ (2008) Inspiration, images and design: an investigation of designers' information gathering strategies. J Design Res 7(4):331–351
11. Nagai Y, Taura T, Mukai F (2009) Concept blending and dissimilarity: factors for creative concept generation process. Des Stud 30(6):648–675
12. Pasman G (2003). Designing with precedents. Ph.D thesis (industrial design). TU Delft, The Netherlands, p 224
13. Rosenman M, Gero J (1992) Creativity in design using a prototype approach, in modeling creativity and knowledge-based creative design. In: Gero JS, Maher ML (eds) Lawrence Erlbaum, London, pp 119–145
14. Schön DA (1992) Designing as reflective conversation with the materials of a design situation. Knowl-Based Syst 5(1):3–14
15. Van der Lugt R (2005) How sketching can affect the idea generation process in design group meetings. Des Stud 26(2):101–122

Chapter 6
FuzEmotion: A Backward Kansei Engineering Based Tool for Assessing and Confirming Gender Inclination of Modern Cellular Phones

Hsiang-Hung Hsiao, William Wei-Lin Wang and Xun W. Xu

Abstract Conventional cellular phone companies adopted manufacturer-orientated philosophy in product design, where the cellular phones are designed purely based on the designers' own perceptions. However, this philosophy is no longer effective as the current market of cellular phones is shifting towards more customer-orientated domain. It has evidenced by several recent studies that customers are emotionally connected to the products during decision making. Companies need to be able to discover these feelings and reflect these feelings back to product design process. Kansei Engineering (KE) is a concept that aimed to solve this type of emotionally associated issues by integrating true customers' felling (Kansei) into the design of products. This chapter presented a Backward KE tool named FuzEmotion, which could be used to assess and confirm gender inclination of modern mobile phones. The goal of this tool was to assist cellular phone designers to create products that can truly reflect the needs and feelings from the end users. A new FuzEmotion system was constructed based on one hundred and twenty cellular phone samples selected from five major cellular phone manufactures (Nokia, Samsung, Sony Ericsson, Motorola and LG),which achieved an overall accuracy of 84.9 %. A database consisted of three hundred and thirty-one cellular phones was an essential part of the FuzEmotion system. The concept behind FuzEmotion and KE is relatively similar, which both aimed to assess end users' feeling and try to integrate these feelings (Kansei) back to the products. Although these two systems are not directly connected to each other, this study has, in part, witnessed the feasibility of KE applications in product design.

H.-H. Hsiao (✉) · W. W.-L. Wang · X. W. Xu
Department of Mechanical Engineering, School of Engineering, The University
of Auckland, Auckland, New Zealand
e-mail: dean_hsiao@hotmail.com

W. W.-L Wang
e-mail: williamw.nz@gmail.com

X. W. Xu
e-mail: xun.xu@auckland.ac.nz

S. Fukuda (ed.), *Emotional Engineering vol. 2*, DOI: 10.1007/978-1-4471-4984-2_6,   73
© Springer-Verlag London 2013

6.1 Introduction

Conventional product designs are mostly driven by the perspective of manufactures. Products are designed purely based on market research or designers' own perceptions. Although advance in technology has enabled the time taken for these design processes to be significantly reduced and the final products are often expected to meet market demands. However, it is questionable that how many of the products may truly reflect the expectations from target customers. Furthermore, the functionality of the products is indeed one of the key parameters that influence customer perceptions during purchasing. However, the trend of functionality focused product design process is believed to decline and gradually be replaced by customer focused approach. In the other word, products should be designed based on customers' perceptions and their true feeling about the products. Kansei Engineering (KE) or aka Emotional Engineering system

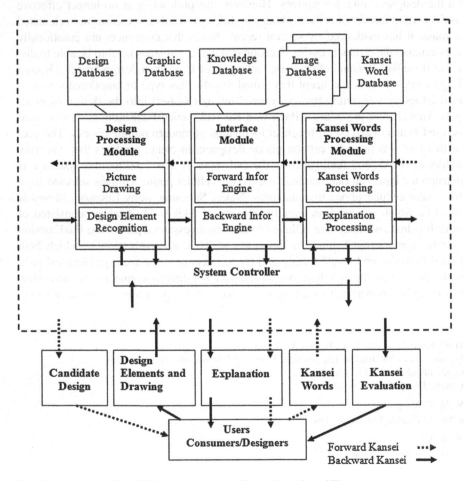

Fig. 6.1 Structure of hybrid Kansei engineering (Reproduced from [5])

is the first system that was developed based on customer-orientated product design philosophy [1–4]. Since initially introduced by Professor Nagamachi in 1970s, KE system has been integrated into several commercial products and tremendous market successes have been achieved. The word Kansei is a Japanese term used to represent feelings or emotions and this element is expected to be the most influential parameters in product design. Professor Nagamachi believed that it was feasible to integrate human emotions into product design processes and enable the product to truly reveal the needs of customers.

As proposed by Professor Nagamachi, the original framework of KE system could be further expended into a new system named Hybrid KE system, which is a decision support system (DSS) for both customers and designers. The Hybrid KE system consisted of two individual systems being Forward KE system and Backward KE systems. In the forward KE system, customer needs are input into KE system via Kansei words. These inputs are then processed and matched with the Kansei words stored in the database. Products that best-fitted customer' requirements or Kansei words are then displayed to the customers, hence helping them to make meaningful decisions during purchasing. On the opposite side, Backwards KE system allowed the designers to assessing products that are already on the markets and translate product features back to Kansei words via KE systems. This concept is very similar to market research, where customers' perceptions on a particular product are recorded and fed back to product design cycle for product improvements. It is also noticeable that both Forward and Backward KE systems adopt customer-orientated product design philosophy, which is believed to be the future trend of product design. A schematic diagram of Hybrid KE System is illustrated as Fig. 6.1.

Previously, cellular phones were used to examine the feasibility of KE applications via the development of FuzEmotion, which was a Backward KE tool that could be used on accessing and confirming the gender inclination of cellular phones [6]. There were two reasons for selecting cellular phones for the study of KE. Firstly, the market for cellular phone is relatively matured than the markets for rest of electronic products. Thanks to advance in technology, a new cellular phone is pushed into the market almost every month with new functionalities. There is no doubt that the functionality of cellular phones has always been the primary considerations when the customers deicide which phone to purchase. However, due to the nature of short shelf-life of cellular phones, it is questionable what factors are actually gaining customers' interests. Shorten product life cycle may be a good sign for cellular phone manufactures, but may not be true if this was looking from the customers' perspective. Customers now have significantly reduced time for decision making during purchasing and functionality is definitely not the only consideration, hence making cellar phones the best product for examine the feasibility of KE Applications.

Secondly, young generations of cellular phone users are not only demanding functionalities, but also style of cellular phones. Unique features of cellular phone that can be used to represent both personality and social status of end users. It is common that most of the conventional cellular phone manufactures adopted the manufacturer-orientated product design philosophy. As the market is currently shifting from manufacturer-orientated product design to customer-orientated product design, cellular phone companies and manufacture need to consider what actually make a particular phone more appealing to a target group. Asking questions such as "What are the key features of cellular phones that may be emotionally connected to end users and response to the Kansei of customers?" This is relevant to the concept of Backward KE to a certain extent. Details about new FuzEmotion system and its development methodologies are discussed in depth in the following sections of report.

This project aimed to improve the overall accuracy of previously developed Fuzzy Logic based FuzEmotion, which is a backward KE tool that may be used on accessing and confirming the gender inclination of modern cellular phones. In addition to the first objective, the project also aimed to explore the feasibility of KE applications on product designs and during product design processes.

6.2 Mobile Phone Database

One of the key building blocks of KE system is product database, which usually consisted of Kansei effective features (attributes) and specifications. The customer perceptions of these features are believed to evoke customer desire on purchasing the products. The overall effectiveness and accuracy of KE applied system is significantly depended on how well the database was constructed. In this project, a new database containing three hundred and thirty-one modern cellular phone samples was constructed to assist the development of FuzEmotion. Several improvements

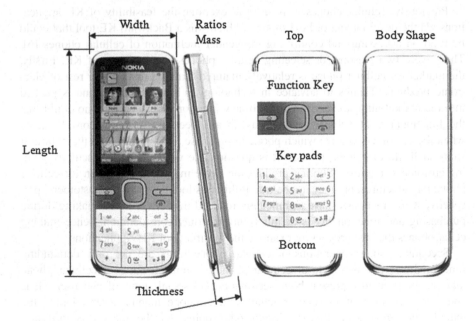

Fig. 6.2 New cellular phone attributes [6]

have been made to the previous database and additional features have been added. For example, the overall sample size for the new database was increased from 229 to 331 and the quantity of cellular phone features (attributes) were increased from 10 to 38, namely dimensions, ratios, mass, top, bottom curvatures etc.

Features that formed the overall shape and appearance (external cover design) of the cellular phones were the only main concern here. For example, the overall shape of the cellular phone, top and bottom curvatures, main keypad arrangements and lastly the shape of function keys.

These features formed the fundamental structure of modern cellular phones, which would be firstly appeared to customers during initial contact, hence were considered to have the most significant influence on customer perceptions of the product. A schematic diagram of previous and new cellular phone attributes is illustrated as Fig. 6.2. Some of the features were extracted using Canny edge detection [7], which will be discussed in the next section.

Features such as keypad function key, top and bottom curvatures and body shape were extracted from existing cellular phone images published online. These images were further processed using Matlab inbuilt function named Canny edge detection, which was developed by Canny in [7]. Differ from conventional edge detection methods that may be fooled by noises, Canny edge detection aimed to optimize the overall quality of edges detection by using two thresholds (detection of both weak and strong edges). The output file will only contains the weak edges if these edges are connected to strong edges. This method enabled errors generated during detection to be significantly minimized and most likely to produce true weak edges. Since some of the cellular phone images used in this project was extracted from multiple online sources, the overall quality of these images may not be consistent. Canny edge detections offered the most effective and accuracy results among all

Fig. 6.3 Cellular phone image processed using canny edge detection [7]

other edge detection methods, hence were adopted to process all the cellular phone images in the database. A schematic diagram of cellular phone images processed using Canny edge detection is illustrated in Fig. 6.3.

Total of 120 gender inclined cellular phones were selected to develop FuzEmotion system, being 40 female phones, 50 gender neutral phones and 30 male phones. Rating and reviews collected from online source were used to determine the relationship between selected cellular phones and targeted audiences (i.e. gender groups). As mentioned before, it was possible that these ratings and reviews are subjective, which made the classification difficult. However, it is reasonable to state that these reviews have fully or at least partially revealed the actual perceptions of the end users and driven by the thoughts of end users. These thoughts and feelings appeared to be emotionally connected to the cellular phones. Human perceptions are often subjective and tended to driven by a particular factor such as sense or experiences, thus an artificial intelligence system such as Fuzzy Logic that can mimic the function of human brains is strongly required for the establishment of scientific and systematic based research.

6.3 Selection of Fuzzy Logic Method

Fuzzy Logic has been selected to process cellular phone attributes and assign these attributes into different groups such as ratios, sizes and forms. Fuzzy logic was selected due to two main reasons. Firstly, since the classifications of cellular phone features were conducted based on customer reviews and ratings collected from online source, due to the nature of these reviews, there is no doubt that some of the classification may be subjective and it is difficult to examine the accuracy of these classifications. However, it is also true that human languages or Kansei words naturally contains different degree of vagueness and commonly harder to be defined quantitatively using mathematical model. Certain degree of freedom must be allowed in the system to process these qualitative data. Fuzzy logic is multivalued logic with degree of membership ranging from zero to one inclusively, where true and false are represented by one and zero respectively [8]. In the other word, qualitative data is no longer given an absolute answer of true or false, but an approximated answer that allows small portion of uncertainties with reasonable accuracy. The nature of Fuzzy Logic allowed this method to be the best solution to deal abstractive problems proposed in this chapter.

Secondly, noted that both Forward KE and Backward KE systems are used to assess human feeling and thoughts about a particular product. These processed feelings are turned into Kansei words and inputted into KE system for product design process improvement. In the other term, a portion of KE systems is trying to mimic the function of human brains or to "see" the product from the customers' perspective. Fuzzy Logic was developed as part of artificial intelligence systems that aimed to perform similar functions, which is also relevant to the concept of KE system. Although it may argued that Fuzzy Logic and KE systems are totally independent from each other, it is believed that functions of Fuzzy Logic can

Fig. 6.4   Five steps of fuzzy logic applications [8]

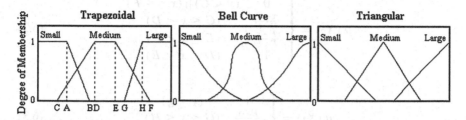

Fig. 6.5   Three typical membership function plots [8, 9]

also be utilized to examine the feasibility of KE systems. Since FuzEmotion was developed based on Fuzzy Logic, FuzEmotion has inherited most of methodologies in Fuzzy Logic systems. The only difference is that FuzEmotion is a tool that was specifically developed to assess cellular phones and confirm the gender inclination on these cellular phones. The methodologies and approaches taken to develop FuzEmotion are discussed in depth in the following sections.

The development of FuzEmotion was based on five individual steps of Fuzzy Logic applications (As indicated as Fig. 6.4).Firstly, Fuzzy inputs (Cellular phone attributes) used to feed into FuzEmotion system must be pre-processed and the degree of belonging of these Fuzzy inputs is determined via membership functions. Membership functions can also be referring to generalized characteristic functions that represent the degree of truth in a Fuzzy Logic system [8, 9]. It is common in Fuzzy Logic systems to represent these numerical membership functions graphically for ease of understanding. Graphical membership functions are termed membership function plots (MFP), which has degree of membership on the y axis and normalized product feature values on x axis.

Membership functions plots can be represented based different shapes, the most common ones are trapezoidal, bell curve and triangular. Application of each shape is significantly depended on the type of data been processed. Since the overall accuracy of the final system is influenced by different type of MFP shape, selection must be done with care. The three typical MFP are illustrated in Fig. 6.5.

Note that each membership function plot is assigned with three subgroups (i.e. Small, Medium and Large). Each subgroup are represented graphically based on numerical membership functions (generalized characteristic functions). The three memberships functions used to define the shape of three subgroups in a membership function plot are listed as Eqs (6.1)–(6.3). English alphabets used in the membership functions are there to represent normalized feature values.

$$\mu(X_1) = \begin{cases} 0 & (x > B) \\ \frac{B-x}{B-A} & (A \leq x \leq B) \\ 1 & (x < A) \end{cases} \tag{6.1}$$

$$\mu(X_2) = \begin{cases} 0 & (x < C) \,||\, (x \leq> F) \\ \frac{x-C}{D-C} & (C \leq x < D) \\ \frac{F-x}{F-E} & (E \leq x < F) \\ 1 & (D < x < E) \end{cases} \tag{6.2}$$

$$\mu(X_3) = \begin{cases} 1 & (x < g) \\ \frac{H-x}{H-G} & (G \leq x \leq H) \\ 0 & (x > H) \end{cases} \tag{6.3}$$

Secondly, applying Fuzzy Logic operators and Fuzzy rules to the MFP established in the previous stage. Typically, Fuzzy rules are in the form of IF–THEN rules, which should be structured as "if the antecedent is true, then the consequent must be true". The total number of Fuzzy rules required to construct the final system is depended on the number of antecedents established and the subgroups assigned for each antecedent. For example, if two antecedents were established to represent two major features of "Ratio" and "Size", and each of the features was assigned into three subgroups such as Low, Medium and High for "Ratio" and Small, Medium and Large for "Size". This has enabled the formation of nine Fuzzy Rules in total for constructing the final FuzEmotion system. In the case of more than one antecedent in the system, Fuzzy Logic Operators are applied to determine the combined firing strength of each rule established.

Thirdly, it is common that multiple Fuzzy rules may be fired at the same time and determination of the output for each set of fuzzy rules and MFP are achieved by applying the implication methods. This method is also defined as the shaping (generation) of consequent from the antecedents. As suggested by Zadeh [8], several different logic operators such as AND (∩) and OR (∪) can be used to apply implications method. For the purpose of this study, AND (∩) was determined to best suit the scenario encountered and used in all Fuzzy Rules defined.

Aggregation is fourth step where all the output MFP generated using different Fuzzy Rules are combined to form final membership function plot, where the maximum of the individual output MFP are summarized into single final MFP (Indicated as Fig. 6.6).

Finally, the final membership function plot generated using aggregation method must be defuzzified into numerical values. It is common in Fuzzy Logic system that different methods such a centre of gravity and centre of maxima are used to defuzzify the final MFP. For simplicity, the method of centre of gravity or aka the centroid method is used for this type of applications, where the centre of the area under the graph is calculated and translated into single crisp value. This crisp value will indicate the location of final consequent (i.e. the phone being "Male", "Gender

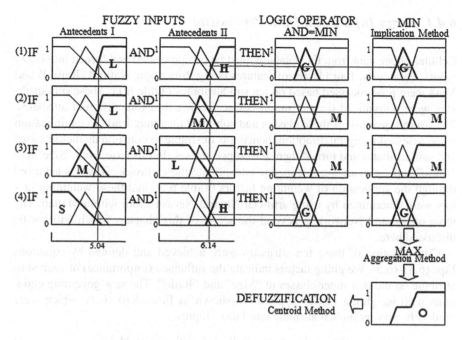

Fig. 6.6 Structure of fuzzy logic system [8]

Neutral" or "Female"). It is also not uncommon for the final crisp value to lie under multiple curves. In this case, the objective being studied may be considered to belong to (intended for) more than one target groups. A schematic diagram to summarize the structure of Fuzzy Logic System is illustrated in Fig. 6.6.

As indicated by Fig. 6.6, normalized feature values are acting as the input to the Fuzzy Logic system and fuzzified via the construction of membership function plot. Each fuzzified MFP is regarded as the antecedent of the system and multiple antecedents are fired through the applications of logic operators, Fuzzy rules and implication method. Once the output of each MFP and Fuzzy rule is generated, aggregation method then combined all the outputs based on the principle of maximum and turned these outputs into single final MFP. The final MFP is then defuzzified via application of centroid method. This is how qualitative data can be translated into final quantitative crisp value, hence indicated the belongings of the object being studied.

6.4 FuzEmotion

As mentioned earlier in the report, a tool named FuzEmotion was developed to examine the feasibility of KE applications via the studies of modern cellular phone features. This section will discuss how the FuzEmotion system was constructed and how these cellular phone features are accessed using such system to confirm the gender inclinations of cellular phones.

6.4.1 Fuzzy Inputs and Data Processing

Cellular phone data from the database must be pre-processed and turned into fuzzy inputs. Previously, four classes of features being dimension, Ratio A, Ratio B and Mass were pre-processed based on ten sub-attributes (Table 6.1). These sub-attributes were normalized (expect mass) before turning into four different attributes. The overall quantity of these classes and sub-attributes may leads to complication when assessed using FuzEmotion, hence it is desirable to consolidate these classes and sub-attributes and turned them into super-classes. In this case, both "Size" and "Ratio" super-classes were defined to consolidate these classes. This was achieved through the application of weighting factors (Table 6.2), and these weighting factors were determined by using artificial Neutral Network tool within Matlab. The topic on Neutral Network is beyond the scope of this chapter, which will not be discussed here.

Consolidation of these ten attributes were achieved and defined by equations Eqs. (6.4)–(6.5). Weighing factors indicate the influence (importance) of each sub-attribute on the two super classes of "Size" and "Ratio". The new governing equations with weighting factors applied are shown as Eqs. (6.6)–(6.7), which were used to fuzzify individual attribute into Fuzzy inputs.

$$Size = l \times L + w \times W + t \times T + m \times M \qquad (6.4)$$

$$Ratio = \alpha \times R_a + \beta \times R_b + \gamma \times R_c + \delta \times R_d + \varepsilon \times R_e + \zeta \times R_f \qquad (6.5)$$

$$Size = 0.1L + 0.2W + 0.7T + 0.5M \qquad (6.6)$$

$$Ratio = 0.2R_a + 0.4R_b + 0.4R_c + 0.4R_d + 0.3R_e + 0.3R_f \qquad (6.7)$$

Similarly, the newly added sub-attributes (Table 6.2) were also pre-processed before fuzzified via membership functions. The external cover design of cellular phones were turned into five different classes of top, bottom, keypads, function key and lastly body shapes. Each class of attribute is associated with five

Table 6.1 Four classes of sub-attributes (Hsiao et al. 2009)

Classes	Sub-attributes	Weighting factors
Dimension	Length (L)	$l = 0.1$
	Width (W)	$w = 0.2$
	Thickness (T)	$t = 0.7$
Ratio A	$R_a = W/L$	$\alpha = 0.2$
	$R_b = T/W$	$\beta = 0.4$
	$R_c = T/L$	$\gamma = 0.4$
Ratio B	$R_d = L/W$	$\delta = 0.4$
	$R_e = W/T$	$\varepsilon = 0.3$
	$R_f = L/T$	$\zeta = 0.3$
Mass	M	$m = 0.5$

Table 6.2 Five new classes and 28 new attributes

Class	Sub-attributes	Weighting (Female)	Weighting (Gender neutral)	Weighting (Male)
Top	Top1	TF1 = 3.1	TG1 = 2.2	TM1 = 2.3
	Top2	TF2 = 0.5	TG2 = 0.4	TM2 = 0.6
	Top3	TF3 = 0.5	TG3 = 0.2	TM3 = 1.3
	Top4	TF4 = 3.6	TG4 = 4.4	TM4 = 2.6
	Top5	TF5 = 1.8	TG5 = 2.4	TM5 = 3.2
	Top6	TF6 = 0.5	TG6 = 0.4	TM6 = 0.0
Bottom	Bottom1	BF1 = 2.3	BG1 = 1.8	BM1 = 2.6
	Bottom2	BF2 = 0.8	BG2 = 0.4	BM2 = 0.6
	Bottom3	BF3 = 0.5	BG3 = 0.6	BM3 = 1.0
	Bottom4	BF4 = 3.8	BG4 = 3.8	BM4 = 2.9
	Bottom5	BF5 = 1.3	BG5 = 2.0	BM5 = 2.9
	Bottom6	BF6 = 1.3	BG6 = 1.4	BM6 = 0.0
Keypad	Keypad1	KF1 = 3.3	KG1 = 4.6	KM1 = 3.2
	Keypad2	KF2 = 2.1	KG2 = 3.2	KM2 = 4.6
	Keypad3	KF3 = 2.0	KG3 = 1.6	KM3 = 1.9
	Kaypad4	KF4 = 1.8	KG4 = 0.6	KM4 = 0.3
	Kaypad5	KF5 = 0.8	KG5 = 0.0	KM5 = 0.0
Function Key	FuncKey1	FF1 = 2.1	FG1 = 2.2	FM1 = 3.5
	FuncKey2	FF2 = 1.8	FG2 = 2.8	FM2 = 1.9
	FuncKey3	FF3 = 3.8	FG3 = 2.6	FM3 = 2.4
	FuncKey4	FF4 = 2.0	FG4 = 0.6	FM4 = 1.3
	FuncKey5	FF5 = 0.3	FG5 = 1.8	FM5 = 0.6
	FuncKey6	FF6 = 0.0	FG6 = 0.0	FM6 = 0.3
Body Shape	Shape1	SF1 = 2.6	SG1 = 3.2	SM1 = 3.2
	Shape2	SF2 = 4.4	SG2 = 5.2	SM2 = 4.9
	Shape3	SF3 = 2.3	SG3 = 0.0	SM3 = 0.0
	Shape4	SF4 = 0.3	SG4 = 0.0	SM4 = 0.0
	Shape5	SF5 = 0.4	SG5 = 1.6	SM5 = 1.9

of six different sub-attributes (Top1, Top2 etc.). As shown in the table, different weighting factors were assigned for each sub-attribute and may not be the same for different gender groups. This was due to the facts that even the same sub-attributes may have different influence on the classes for different gender groups. For example, sub-attribute Top1 is more important for the female "Top" class (w.f. of 3.1) than for the male "Top" class (w.f. of 2.3). These weighting factors were calculated based on degree of appearance or fluctuations of sub-attributes. For example, as a particular sub-attribute is adopted much more than rest of the sub-attributes for designing new cellular phone (i.e. female phone), it is reasonable to state that such sub-attribute has more influence on the feature design of that particular cellular phone and make the design more attractive to female. Note that the weighting factors were multiplied by a factor of 10 to allow the new MFP to have similar scale as the other two MFPs (Size and Ratio). So the new MFP would not contract as much relative to the other two MFPs.

Table 6.3 Super class "Form" and sub-classes of "Form"

Super-class	Sub-classes	Weighing factors
Form	Top	T = 0.19
	Bottom	B = 0.17
	Keypad	K = 0.23
	Function Key	FK = 0.23
	Body Shape	BS = 0.18

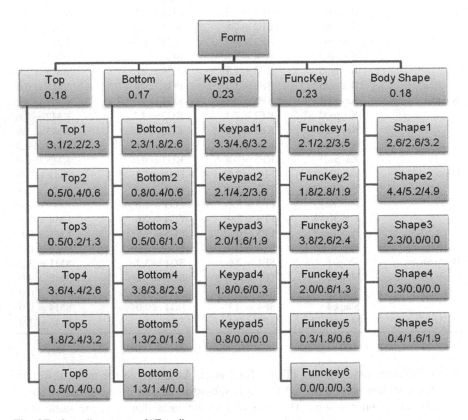

Fig. 6.7 Overall structure of "Form"

Due to complexity of these classes and sub-attributes, consolidation was conducted once again to unify these classes into single super-class. The new-super class was given name as "Form", which defines the external cover design of cellular phones. The new super-class of form and its sub-classes are listed in Table 6.3.

As shown in Table 6.3 that new weighting factors were applied again on each sub-classes of "Form". These weighting factors were also calculated based on the degree of appearance or fluctuations of each sub-class. This was looking at the overall influence of individual class (Top, Bottom etc.) on the super-class of

"Form". The overall structure of this process can be summarized by Fig. 6.7 with relative weighing factors underneath each sub-attributes and classes, where the weighting factors are in the order of female, gender neutral and male.

Note that there are some zero weighting factors such as the weighing factors for Keyad5 and Funckey6, meaning that these sub-attributes were not adopted at all for a particular class and should not have influence on the form design regardless of any gender groups. The new super-class of "Form" is now defined by Eqs. (6.8) and (6.9).

$$Form = T \times Top + B \times Bottom + K \times KekPad + FK \times FunctionKey + BS \times BodyShape \tag{6.8}$$

$$Form = 0.18Top + 0.17Bottom + 0.23KekPad + 0.23FunctionKey + 0.18BodyShape \tag{6.9}$$

All the three sets of Fuzzy inputs were then fuzzified via membership functions and represented graphically with MFPs. These MFPs would then act as the antecedents in the FuzEmotion system and combined by Logic Operators, Fuzzy Rules and implication method. The formation of individual MFP is discussed in the following sections.

6.4.2 Membership Functions

In this project, three different sets of antecedents were defined based the study of cellular phone feature data and these antecedents were acted as the input of FuzEmotion system. Each set of antecedents are assigned with three sub-categories, where each of the sub-categories of data was fuzzified by one membership function. For "Size" feature, the sub-categories were assigned as "Small", "Medium" and "Large". Similarly for "Ratio" feature, "Low", "Medium" and "High" sub—categories were assigned. Lastly for "Form" feature, the sub-categories were assigned as "Form 1", "Form 2" and "Form 3". Each of these sub-categories was associated with a membership function, which used to transform these sub-categories into MFP. The membership functions for each sub-category are listed as Eqs. (6.10)–(6.18).

$$(S_1) = \begin{cases} 0 & (x > 5.33) \\ \frac{5.33-x}{0.79} & (4.54 \leq x \leq 5.33) \\ 1 & (x < 4.54) \end{cases} \tag{6.10}$$

$$\mu(S_2) = \begin{cases} 0 & (x < 4.24)\,or\,(x > 5.71) \\ \frac{x-4.24}{0.57} & (4.24 \leq x < 4.81) \\ \frac{5.71-x}{0.90} & (4.81 \leq x < 5.71) \\ 1 & (x = 4.18) \end{cases} \tag{6.11}$$

$$\mu(S_3) = \begin{cases} 0 & (x < 4.94) \\ \frac{5.44-x}{0.50} & (4.94 \leq x \leq 5.44) \\ 1 & (x > 5.44) \end{cases} \quad (6.12)$$

$$\mu(R_1) = \begin{cases} 0 & (x > 6.19) \\ \frac{6.19-x}{0.14} & (6.05 \leq x \leq 6.19) \\ 1 & (x < 6.05) \end{cases} \quad (6.13)$$

$$\mu(R_2) = \begin{cases} 0 & (x < 6.00) \, or \, (x > 6.30) \\ \frac{x-6.02}{0.11} & (6.02 \leq x < 6.13) \\ \frac{6.30-x}{0.17} & (6.13 \leq x < 6.30) \\ 1 & (x = 6.13) \end{cases} \quad (6.14)$$

$$\mu(R_3) = \begin{cases} 0 & (x < 6.05) \\ \frac{6.21-x}{0.16} & (6.05 \leq x \leq 6.21) \\ 1 & (x > 6.21) \end{cases} \quad (6.15)$$

$$(F_1) = \begin{cases} 0 & (x > 34.8) \\ \frac{34.8-x}{8.5} & (26.3 \leq x \leq 34.8) \\ 1 & (x < 26.3) \end{cases} \quad (6.16)$$

$$\mu(F_2) = \begin{cases} 0 & (x < 20.1) \, or \, (x > 35.9) \\ \frac{x-20.1}{8.4} & (20.1 \leq x < 28.5) \\ \frac{35.9-x}{7.4} & (28.5 \leq x < 35.9) \\ 1 & (x = 28.5) \end{cases} \quad (6.17)$$

$$\mu(F_3) = \begin{cases} 0 & (x < 20.8) \\ \frac{31-x}{10.2} & (20.8 \leq x \leq 31) \\ 1 & (x > 31) \end{cases} \quad (6.18)$$

The key parameters for each antecedents generated are presented in Table 6.4 and the respective membership function plot are illustrated as Figs. 6.8, 6.9 and 6.10. These MFP are used as the input of FuzEmotion systems.

MFP are used to represent fuzzified cellular phone data via membership function, which presented as a specific area in the MFP. Please note the values on the y axis of the MFP. As mentioned earlier, the value ranged from zero to one (inclusively) on the y axis is used to indicate the degree of membership (aka degree of belongings) of fuzzified data. Taking "Size" MFP as an example, a fuzzified size

Table 6.4 Membership function parameters

Fuzzy set	A	B	C	D	E	F	G	H
Size	4.54	5.33	4.24	4.81	4.81	5.71	4.98	5.44
Ratio	6.05	6.19	6.02	6.13	6.13	6.30	6.05	6.21
Form	26.3	34.8	20.1	28.5	28.5	35.9	20.8	31

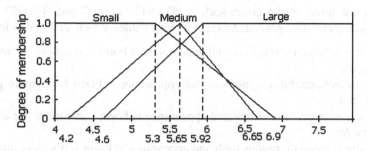

Fig. 6.8 Membership function plot for "Size" (Hsiao et al. 2009)

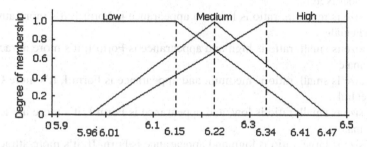

Fig. 6.9 Membership function plot for "Ratio" (Hsiao et al. 2009)

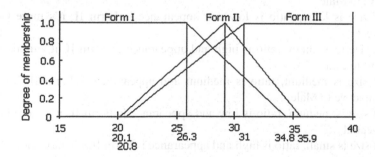

Fig. 6.10 Membership function plot for "Form"

feature value of 5.3 or less are stated to be fully belonged to the sub-category of "Small" with a degree of membership of 1, hence 100 % degree of belonging.

6.4.3 *Fuzzy Rules and Logic Operators*

Development of Fuzzy Rules is one of the key steps in the construction of FuzEmotion. The number of Fuzzy Rules required is related to the total number of antecedents and the sub-attributes used in the system. In this case, three antecedents were established being "Size", "Ratio" and "Form". Each of the antecedents was assigned with three sub-categories such as "Form1", "Form2" and "Form3". These nine sub-categories have generated total of 27 Fuzzy rules, which are listed as follows:

R1 If size is large, ratio is high and appearance is Form I, it's more attractive to Female

R2 If size is large, ratio is medium and appearance is Form I, it's more gender Neutral

R3 If size is large, ratio is low and appearance is Form I, it's more attractive to Female

R4 If size is medium, ratio is high and appearance is Form I, it's more attractive to Female

R5 If size is medium, ratio is medium and appearance is Form I, it's more Gender Neutral

R6 If size is medium, ratio is low and appearance is Form I, it's more attractive to Female

R7 If size is small, ratio is high and appearance is Form I, it's more attractive to Female

R8 If size is small, ratio is medium, and appearance is Form I, it's more Gender Neutral

R9 If size is small, ratio is low, and appearance is Form I, it's more attractive to Female

R10 If size is large, ratio is high and appearance is Form II, it's more attractive to Male

R11 If size is large, ratio is medium and appearance is Form II, it's more attractive to Male

R12 If size is large, ratio is low and appearance is Form II, it's more Gender Neutral

R13 If size is medium, ratio is high and appearance is Form II, it's more attractive to Male

R14 If size is medium, ratio is medium and appearance is Form II, it's more attractive to Male

R15 If size is medium, ratio is low and appearance is Form II, it's more Gender Neutral

R16 If size is small, ratio is high and appearance is Form II, it's more attractive to Male

R17 If size is small, ratio is medium and appearance is Form II, it's more attractive to Male

R18 If size is small, ratio is low and appearance is Form II, it's more Gender Neutral

R19 If size is large, ratio is high and appearance is Form III, it's more Gender Neutral

R20 If size is large, ratio is medium and appearance is Form III, it's more Gender Neutral

R21 If size is large, ratio is low and appearance is Form III, it's more Gender Neutral

R22 If size is medium, ratio is high and appearance is Form III, it's more Gender Neutral

R23 If size is medium, ratio is medium and appearance is Form III, it's more Gender Neutral

R24 If size is medium, ratio is low and appearance is Form III, it's more Gender Neutral

R25 If size is small, ratio is high and appearance is Form III, it's more Gender Neutral

R26 If size is small, ratio is medium and appearance is Form III, it's more Gender Neutral

R27 If size is small, ratio is low and appearance is Form III, it's more Gender Neutral

These Fuzzy Rules are determined based on the studies of cellular phone data and the antecedents generated previously. As stated by Zadeh [8], Logic operators [In this case AND ($\cap$)] are used to activate Fuzzy Rules.

U_S, U_R and U_F are used to represent the resulting degree of membership from the antecedents of "Size", "Ratio" and "Form" respectively.

$$\text{AND} : U_{S \cap R} = U_S \wedge U_R \wedge U_F = \min\,(U_S, U_R, U_F) \qquad (6.19)$$

6.4.4 Implication Method

Implication method was applied to determine the final output from each set of MFP. It is a process involving application of Logic Operator and determining the minimum of result from each antecedent fired. Since there are total of 27 sets of MFPs and Fuzzy Rules, 27 sets of output will be generated due to the firing of these MFPs. It is difficult to present all 27 sets MFPs here due to space limitation, but a simple illustration of single implication method is presented in Fig. 6.11 (Based on Rule 13).

Note that the resulting degree of membership is 0.52 for "Size", 0.6 for "Ratio" and lastly 0.65 for "Form". Since AND ($\cap$) logic operator is applied alongside of implication method, the minimum of the three resulting degree of membership is taken as the output for this particular set of MFPs (Fuzzy Rule 13). It is confident to state that for this particular set of MFPs, the cellular phone being study has been classified into Male group with 52 % degree of membership. This can be represented by Eq. (6.20):

$$U_{S \cap R} = U_S \wedge U_R \wedge U_F = 0.52 \wedge 0.6 \wedge 0.65 = 0.52 \qquad (6.20)$$

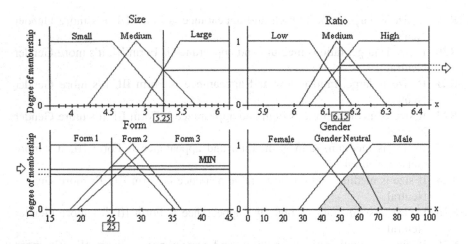

Fig. 6.11 Implication method conducted based on rule 13

The final shape of gender MFP was initially approximated. Since all the antecedents MFPs (Ratio, Size and Form) were established based three membership functions and taken the shape of trapezoidal, it is reasonable to assume the final membership function plot must have adopted similar kind of shape.

6.4.5 Aggregation and Defuzzifications

Aggregation was a process used to unify all the outputs from each set of MFPs into single final MFP or aka final membership function plot. The OR (∪) logic operator was used to determine the maximum of the outputs. Since all the Fuzzy rules are fired in parallel, it likely to have similar output from each set of MFPs but with different degree of membership. Note that both AND (∩) and OR (∪) are used in the whole construction process of FuzEmotion, which were the key elements in FuzEmotion system.

Defuzzification is the last step in the construction of FuzEmotion, where the final MFP was defuzzified into single crisp value. This value can be used by cellular phone designer to assess the gender inclination of mobile phones. As mentioned earlier, several methods could be adopted to defuzzify the final MFPs such as centre of mass or centre of maxima. For simplicity, centre of mass is applied to deal the problem presented in this project. Centre of mass method worked upon determining the centroid of the area under the curves. All 120 cellular phone have been defuzzified into final crisp values. The overall accuracy of previous FuzEmotion system and the current FuzEmotion system are presented in Table 6.5.

Table 6.5 Overall accuracy of FuzEmotion system

Outcome	Accuracy(Old) (%)	Accuracy(New) (%)
Female phone correctly identified as female	74.0	87
Female phone mistakenly identified as male	13.3	6
Female phone mistakenly identified as gender-neutral	12.7	7
Gender-neutral phone correctly identified as gender-neutral	80.0	92
Gender-neutral phone mistakenly identified as male	2.0	2
Gender-neutral phone mistakenly identified as female	18.0	6
Male phone correctly identified as male	73.0	75.6
Male phone mistakenly identified as gender-neutral	19.0	20
Male phone mistakenly identified as female	8.0	4.4
Overall accuracy	76.0	84.9

6.5 Recap and Evaluation

The process taken to improve the overall accuracy of FuzEmotion system is an iterative process, where the reviews and rating from end users, cellular phone data from the manufacturers and particularly the weighting factors of each sub-attributes are used to "tuning" the overall system. It could be expected that upon "tuning", the accuracy of FuzEmotion system could be improved. However, due to the nature of Fuzzy Logic based system, certain degree of uncertainties was allowed in the system, and hence achieving 100 % of system accuracy was not feasible. However, it is also reasonable to state that with 84.9 % of overall system accuracy, FuzEmotion system is still a very effective tool on assessing and confirming the gender inclination of cellular phones. The solution of FuzEmotion system is a final membership function plot that is integrated with 27 different sets of MFPs and Fuzzy Rules. The final MFPs for the previous and current FuzEmotion system are presented as Fig. 6.12.

As indicated by Fig. 6.12, the final MFP for the current system is shifted towards male side and the previous MFP was shifted towards female side. This change may be due to two reasons. Firstly, there are more gender neutral phones in the selected database (i.e. 50), which enabled to overall result to shift more towards the male side. Secondly, it was possible that the most current cellular phone design may be intended for gender neutral groups, hence allowing these phones to gain benefits and attentions from all the three gender groups. Note that these values on the x axis are there only for the ease of calculating the centroid of the result.

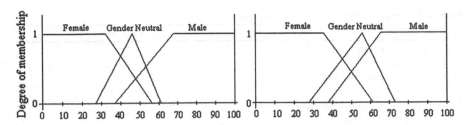

Fig. 6.12 Final membership function plot for previous (*Left*) and current (*Right*) FuzEmotion System (Hsiao et al. 2009)

6.6 Conclusion

Conventional cellular phone manufactures are more focused on the functionality issues of the phones. There is no doubt that functionality of the products has always being one of the important considerations when the end users decide what product to purchase. However, as technologies has pushed the design of cellular phones to the limit with relatively minimised development time. The philosophy of manufacturer-orientated design is no longer effective in the current market of cellular phones. End users are not only satisfied by the functionality of cellular phones, but unique factors such as the appearance of the phone, the style of the phone and lastly good feeling or Kansei about the phone. Cellular phone manufacturers and companies need to look at this issue from a totally different perspective. Questions such as "What are the actual factors that gain the attentions of end users?", "How these products are emotionally connected to the customers?" and lastly "what influences these emotions?" must be answered. KE is a concept that aimed to solve this type of abstractive emotionally associated issues by assessing customers' true perceptions and integrate these assessed data into product design process, hence achieving market success. The needs from customers are satisfied both emotionally and functionally by KE integrated products.

This project focused on the study of cellular phone features and how these features are emotionally connected to end users. This was achieved through the development of FuzEmotion system, which is Backward KE tool that could be used to assess and confirm the gender inclination of mobiles with an overall accuracy of 84.9 %. In the word, FuzEmotion system will be able to tell the users which gender group (female, gender neutral or male) a particular cellular phone belongs to. The whole FuzEmotion system consisted of 27 sets of antecedent MFPs, 27 Fuzzy rules and 1 final membership function plot. The system was constructed based on 120 specifically selected phones from a database containing total of 331 phone samples. These samples were extracted from officially released source published on the World Wide Web. These cellular phones were designed and manufactured by the top five cellular phone companies being Nokia, Samsung, Sony Ericsson, Motorola and lastly LG, which well represented the current cellular phone market.

Throughout the development of FuzEmotion, several key factors were discovered. For example, the weighting factors used to generate MFPs had significant influences on the overall accuracy and Kansei aspects of the system. Note that the current system tended to shift towards male side. This phenomenon may due to two major reasons. Firstly, the system might has witnessed the trend of current cellular phone design as recently released phones are intended for male or gender neutral customers. Secondly, it was also likely that such change was affected by the quantity of gender neutral phones in the database. Regardless of the difference, FuzEmotion has prove its capability to assess and confirm gender inclination of cellular phone and may acted as a design supporting tool for cellular phone designers.

This project has achieved, in part, the two objectives proposed in the earlier sections. The overall accuracy of previous FuzEmotion system has improved from 76 to 84.9 % by adding additional attributes and features into the system. Although FuzEmotion was not directly related to KE, the concepts behind both systems are relatively similar, which both aimed to study the feeling of customers. It is reasonable to state that the application KE in product design process should be a feasible approach based on the study of FuzEmotion conducted. Although KE has been existed for almost 40 years, there is no doubt that significant amount of studies are still required until the function of KE system is fully utilized.

References

1. Nagamachi M (1989) Kansei engineering. Kaibundo Publishing, Tokyo
2. Nagamachi M (1995) Kansei engineering: a new ergonomic customer-orientated technology for product development. Int J Ind Ergon 15:3–11
3. Nagamachi M (1974) A study of emotion-technology. Japan J Ergon 10(4):121–130
4. Nagamachi M (2002) Kansei engineering as a powerful consumer-oriented technology for product development. Appl Ergon 33(3):289–294
5. Matsubara Y, Nagamachi M (1997) Hybrid Kansei engineering system and design support. Int J Ind Ergon 19:81–92
6. Hsiao HH, Wang WW (2009) Emotional feature evaluation of modern cellular phones with Neural Network and Fuzzy Logic. Part IV project report, 2009-ME21, University of Auckland
7. Canny J (1986) A computation approach to edge detection. Trans Pattern Anal Mach Intell 8(6):679–698
8. Zadeh LA (1965) Fuzzy sets. Inf Control 8:338–353
9. Urena R, Rodriguez F, Berenguel M (2001) A machine vision system for seeds germination quality evaluation using Fuzzy Logic. Comput Electron Agric 32:1–20

Chapter 7
ProEmotion: A Tool to Tell Mobile Phone's Gender

William Wei-Lin Wang, Hsiang-Hung Hsiao and Xun W. Xu

Abstract Kansei engineering, also known as Kansei ergonomics or emotional engineering, aims at analyzing and incorporating customer's feeling and demands into product function and product design. The chapter described a system called ProEmotion for the purpose of assessing the Kansei aspects of a product by considering design attributes of a product. Neural Network is used to process Kansei words. The system has been successfully implemented to ascertain gender inclination of a mobile phone. Principal parameters of a mobile are considered, that is, length, width, thickness and mass. The system can inform gender inclination of a mobile phone with accuracy up to 90 %. This is based on a set of 92 mobile phone samples from the five major mobile phone manufacturers.

7.1 Introduction

The trend of product development has become more and more consumer-oriented. As well as consumer's requirement for the needed functionality, consumer's feeling toward a product has also become an important factor in their choice of the product. Kansei engineering, also known as Kansei ergonomics or emotional engineering, aims at effectively analyzing and incorporating customer's feeling and inclination into product function and product design. Kansei engineering was founded in Japan in the late 1970s [1–3].

When a customer sets off to purchase a product, for instance a mobile phone, he/she often gets emotionally connected with the phone before he/she is committed to purchasing [4–6, 7]. An example of emotional connectedness would be

W. W.-L. Wang (✉) · H.-H. Hsiao · X. W. Xu
Department of Mechanical Engineering, School of Engineering, The University of Auckland, Auckland, New Zealand
e-mail: williamw.nz@gmail.com

H.-H Hsiao
e-mail: dean_hsiao@hotmail.com

X. W. Xu
e-mail: xun.xu@auckland.ac.nz

Fig. 7.1 Hybrid Kansei
engineering system

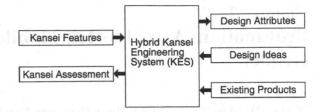

"graceful and looks cute, but not so expensive." This feeling is called "Kansei" in Japanese. In other words, Kansei means the customer's psychological feeling as well as embracing physiological issues [8]. Kansei words are the expressions of the attributes of Kansei aspects.

A Kansei engineering system is often made up of a computer-assisted system of Kansei engineering, an expert system and databases [9–18]. Kansei words are input into the system and recognized in reference to the Kansei word in the databases. The words are matched to the image database and calculated by an inference engine to find the best-fit design, which may be shown for example on the display of a computer. This system is useful for the customer to select a product most fitted to his/her Kansei. It is also very useful to assist a designer to better understand the new product being developed [8]. Such a system supports a flow from Kansei to the design domain and hence is called forward Kansei engineering (forward KE).

A Kansei engineering system may also perform Kansei assessment or evaluation over an existing product or idea in order to recognize Kansei word/features. These Kansei findings can then be used for product improvement or new product development. This is called backward Kansei engineering (backward KE). Often, a Kansei system is based on both forward and backward KE, hence the name hybrid Kansei engineering (hybrid KE) system (Fig. 7.1). Today, applications of Kansei engineering have been found in a number of industries such as automotive, construction, machine tools, electric home, costume, cosmetics [8, 19–28].

Many modern electronic products such as mobile phones have reached their functionality peak for most of the mobile phone users. Consumers are slowly shifting their focus from functionality to fashion. Young people in particular want their phones to be unique, representative of personality and symbol of fashion. Consequently, modern day consumers are a lot harder to be satisfied when it comes to personalized electronic goods. One major effect on mobile phone manufacturers is the demand for an increased pace of new product introduction because the mobile phone's shelf-life has been greatly shortened. A wide range of new mobile phones are being designed and pushed into the market to meet a population of diverse customers.

However, there are few quantitative methods the companies can use to ascertain whether their new models appeal to the targeted consumer. Reviews have indicated that there are mobile phones in the market that were intended toward a specific group of consumer but in fact failed to meet the consumer's expectations [29, 30].

It is evident that modern mobile phone manufacturers are in a matured market in terms of availability, technical functionality and cost, but there is a need to pay greater attention to the Kansei (emotional or affective) aspects of their products. In other words, the companies need to ask questions such as, (a) "what makes a phone more appealing to a male as against a female?" and "(b) what are the design attributes of a mobile phone that contribute to its Kansei?" This is effectively a backward Kansei engineering exercise (Fig. 7.1).

This chapter presents a Neural Network method for appraisal of mobile phone's gender inclination in terms of male, female and general public (i.e., gender-neutral). The product attributes considered are only the principal ones, namely length, width, thickness and weight of a phone. Choice of these attributes is discussed in the following section.

7.2 Mobile Phone Database

A database was first constructed containing detailed information about mobile phones that were on the market in 2009. A total number of 229 modern mobile phones from some of the major brands (i.e., Nokia, Samsung, LG, Motorola and Sony Ericsson) have been collated. The information sourced is that of the officially released product specifications (http://www.sogi.com.tw), including dimensions, mass, keypad ratio, screen ratio, color, price, type (flat, slide, fold or touch) and most importantly targeted customers (male, female or gender-neutral). Figure 7.2 shows the exact dimensions extracted from a phone.

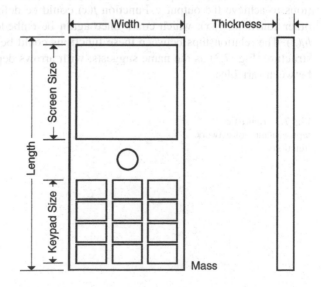

Fig. 7.2 Example of mobile phone attributes

Based on the discussions earlier, only principal attributes, that is, dimensions and mass, are considered in this research. The relative ratios of keypad and screen size to phone length are two attributes that are commonly believed to have a strong bearing on Kansei aspects of a phone. It has, however, been discovered that the majority of 229 mobile phones have very similar keypad/phone-length ratio (0.5) and screen/phone-length ratio (0.3), respectively. This is true across the board of different brands and customer groups. Both functionality and price information are not considered as they have little bearing on a phone's Kansei.

7.3 Selection of Neural Network Method

The nature of the research requires classification of any given mobile phone data into the gender it reflects and is intended for. Neural Network is chosen as a solution toward modeling the problem. The mobile phone database provides ample input data, which is supposedly to be "mapped" to a specific output, that is, gender classification. The key is then to figure out the (explicit) numerical transformation functions in-between, which in many cases may or may not exist. Due to the ability to learn from data like a human brain, one of the advantages of Neural Network is its ability to work with raw data alone without requiring deeper and thorough understanding of the process in the process of obtaining the required result. Other advantages include being robust toward noisy data and its broad application in many abstract classification problems.

In essence, Neural Networks are trained to solve problems based on generating functions to correlate between input data and desired output. The numerical operations behind Neural Network are essentially mathematical models defining a function, $f(x) = y$. Given a particular input x, Neural Network generates functions to achieve the output y. Function $f(x)$ could be defined as a composition of other functions $g_i(x)$, which could once again be embedded into other functions $h_i(x)$. The relationships between these functions could be visualized as a network structure (Fig. 7.3) as the name suggests, with arrows depicting the dependencies between variables.

Fig. 7.3 Heuristic representation of network functions

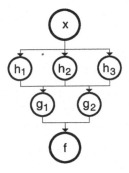

7.4 ProEmotion

To ease the implementation task, many software solutions such as Matlab's Neural Network toolbox have been developed. ProEmotion is a software tool based on the Matlab Neural Network toolbox. This software considers the database of mobile phones and their targeted gender market to give opinions on mobile phones. These opinions are expressed as percentage attractiveness toward the Male and/or

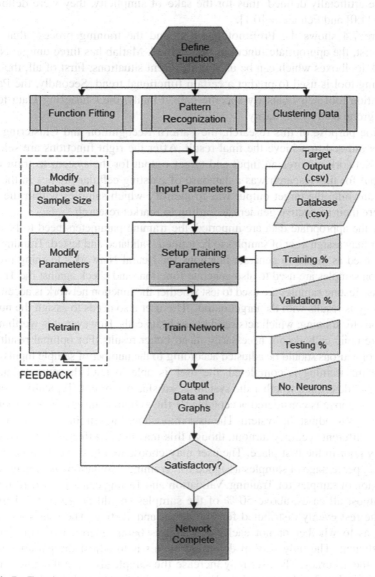

Fig. 7.4 ProEmotion setup and training process

Female audience. ProEmotion can also produce and plot the accuracy of functions, give importance of each parameter, display similar phones to a given phone and save results to a file.

The Comma-Separated Values (CSV) type file format is used as input. Therefore, feature data were listed on a spreadsheet-like format with no labeling for rows and columns. Target audience for each phone is imported into Neural Network as a separate CSV file with corresponding arbitrary values for "Male" and "Female" audience. Any value or vector representing Male and Female could be artificially defined, thus for the sake of simplicity, they were defined as Male = [1.0] and Female = [0.1].

Figure 7.4 shows the ProEmotion setup and the training process that takes place. First, the appropriate functions are defined. Matlab has three unique Neural Network toolboxes which can be used in different situations. First of all, the function-fitting tool is used to predict a certain functional trend. Secondly, the Pattern Recognition tool helps classify specific data. Finally, the Clustering Data tool is used to group data by similarity.

For the purpose of this research, the Pattern Recognition and Clustering Data tool was utilized to achieve the final result. After the right functions are selected, Neural Network requires an input and target output for the system to train upon. This input for this research was a database of existing cellular phones in the market. Meanwhile, the target output was to identify which samples within the database were their respective genders according to market research.

After the appropriate data are imported, the training parameters need to be setup. The user may assign a set of samples to be trained, validated and tested. Training samples are used as means to generate functions for Neural Network's learning process. Validation samples are used to also generate functions and check against the Training functions. Testing samples are used to test whether the function network is accurate by comparing its results with the target output. The user also needs to assign the number of neurons for training, which acts similarly to brain cells. However, it is worth noting that more brain cells do not necessarily mean better results. For optimal results, the number of neurons should be adjusted according to the number of sample inputs.

Once the training is completed, the user is able to read the output data and graphs, and determine whether the system is satisfactory or not. Typically, accuracy of 75 % or above is considered acceptable. If the system is unsatisfactory, there are several ways to adjust the system. The user may choose to retrain the network, hoping for a different accuracy turnout, though this may not be effective if the accuracy was very poor in the first place. The user may choose to adjust the training parameters (i.e., percentage of samples assigned for Training, Validation and Testing). The distribution of samples for Training, Validation and Testing can affect the results.

In almost all cases, above 50 % of the samples should be given for Training, while the rest evenly distributed for Validation and Testing. The user should take caution as to whether or not each parameter is being overloaded with samples or insufficient. The only way of determining this is to adjust the parameters and observe the accuracy. The user may increase the sample size, but the new samples must be closely related to the original ones.

7.5 Application of ProEmotion on Mobile Phones

As the purpose of Neural Network is to differentiate between Male and Female phones, during the training stage, it was necessary to remove certain cellular phone samples from the database which possessed similar attractiveness to both Male and Female. These particular samples, defined as "gender-neutral" phones, tend to confuse the system and prevent it from training accurately. Consequently, the final feature database consists of 92 samples. Ten features are used. They are length, width, thickness, width over length, thickness over width, thickness over length, length over width, width over thickness, length over thickness and mass.

By manually adjusting the Neural Network settings for optimal adaptation with the mobile phone feature database, these settings were built into the foundation of ProEmotion. ProEmotion then imports the database CSV and the target CSV to automatically initiate Neural Network's Pattern Recognition toolbox (Fig. 7.4). This toolbox is responsible for training the system to provide output on the attractiveness toward each gender. After training is completed, Pattern Recognition toolbox automatically stores its output and shuts down, which subsequently initiates Neural Network's Cluster Data toolbox. This toolbox attempts to determine the level of influence of each feature toward the customer's emotions, which is achieved by importing the database CSV and grouping each feature into clusters.

7.6 ProEmotion Outputs

ProEmotion can produce four types of outputs. They are percentage attractiveness, accuracy diagram, weighting plane and similar phones. These outputs are discussed in the following section.

The attractiveness percentage output is self-explanatory. The larger the value, the more attractive the phone is to a target customer. Figure 7.5 shows an example.

The accuracy diagram provides information about the reliability of ProEmotion after classification training. ProEmotion obtains this accuracy by testing its own developed logic against the target CSV for every cellular phone sample within the

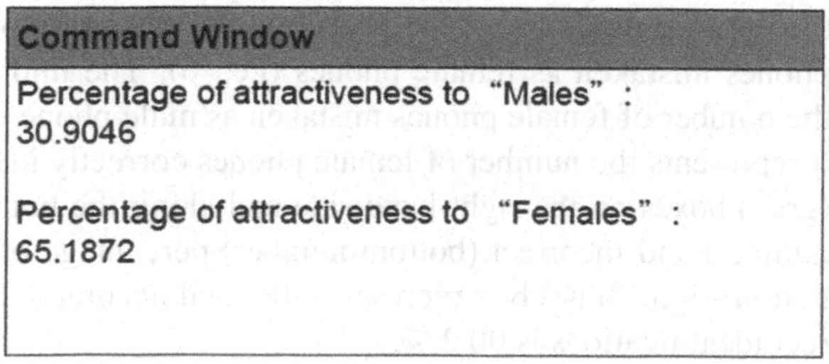

Fig. 7.5 Example of ProEmotion attractiveness output

Command Window

Percentage of attractiveness to "Males" : 30.9046

Percentage of attractiveness to "Females" : 65.1872

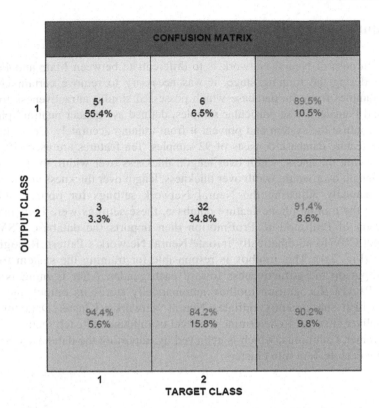

Fig. 7.6 ProEmotion accuracy plot

database. During the training process, 50 % of the samples within database are randomly chosen to be trained, 25 % of samples are used for validation and 25 % are used for testing. Depending on which samples are randomly picked for whichever purpose, the effectiveness and accuracy of the system may vary. Currently, ProEmotion achieves a constant accuracy of 88–90 %, which is considered as much more than adequate for a Neural Network system.

Figure 7.6 shows an example of accuracy plot. The data appear as a 3 × 3 matrix. The top left (green) box of the figure represents the number of male phones correctly identified as male phones (in this case 51, which is 55.4 % of the entire sample size). The (red) box to the right represents the number of male phones mistaken as female phones (i.e., 6). The middle-left (red) box represents the number of female phones mistaken as male phones, and the (green) box next to it represents the number of female phones correctly identified as male phones. The (gray) boxes on the right-hand side and along the bottom tally up the correct (top number) and incorrect (bottom number) percentage of each row and column. The bottom-right (blue) box represents the total accuracy. That is, the accuracy for correct identifications is 90.2 %.

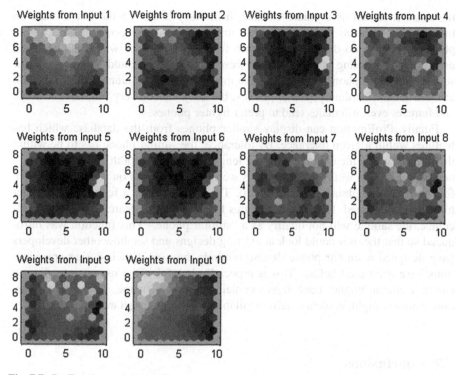

Fig. 7.7 ProEmotion weighting plane

Weighting planes are an effective diagram that can communicate to the user the importance of each feature toward the human emotions. The diagram appears to be a honey-comb structure with ten by ten hexagons (Fig. 7.7); each hexagon is assigned a neuron to think with. A neuron analyzes the samples assigned to it and determines whether certain features have significant effect on the consumer's buying decision or not.

"Weights from Input 1" means the weighting assigned by each neuron for the first feature, that is, length. Then, in the numerical order, there are "weights" for width, thickness, width over length, thickness over width, thickness over length, length over width, width over thickness, length over thickness and mass. The darker each neuron (hexagons) shows, the more important that feature is according to that specific neuron. A general conclusion could be drawn from looking at the overall shade of the one hundred neurons for each feature. For example, it could be said that length is not a major factor for attracting males and females due to the overall bright shade, which by experience makes sense because slide phones and flip phones could get really long once opened, but are still fashionable. Meanwhile, width is more of a concern and thickness nonetheless could be said as the most important factor, this is reasonable as females tend to dislike wider and thicker phones.

The only comment worth mentioning for the ratios (inputs 4 to 9) is that similar shading patterns represent a strong correlation between those features. Lastly, the mass is shown as a gradient shade, from very bright on the left to dark on the right. This means that this feature's importance varies according to different phones. There is no definite explanation to this situation as which models each neuron was analyzing is unknown. However, generally it could be reasoned that some upper-class phones with expensive materials (such as aluminum or crystals) are heavier than ordinary plastic phones, but they can be very attractive to male and females even if females tend to prefer lighter phones.

Finally, ProEmotion can display similar phones from the database which has features within a given tolerance. By tolerance, the similar phone has to be within the following specifications: ±5 mm in length, ±3 mm in width, ±2 mm in thickness and ±20 g in weight. These values were selected based on weighting of each feature (from weighting plane diagram). The more critical a feature, the smaller tolerance is assigned. ProEmotion ensures that if even one feature falls out of tolerance, the sample will not qualify as a "similar phone." This function was introduced so that the user could look at existing designs and see how other developers have designed a similar phone. It also reminds the user if their design specifications have been used before. This is especially helpful if the user is intending to create a unique phone. Each figure contains a picture, name, screen size, type, dimensions, weight, camera quality, available colors and target audience.

7.7 Conclusions

Product functionality has been one of the main focuses for any product development and is responsible for its success in market, too. Appearance, material, shape and form provide the most immediate product data for the developers as well as users. However, less tangible issues such as emotional bonding of users with products, cultural perceptions and social values also provide valuable insights into the product developer to help expand knowledge and understanding of the users need beyond functionalities. Emotional or Kansei engineering is a concept of designing a product to satisfy the emotional needs of its consumers. Unlike marketing and arts, which focuses on bending the customer's emotions toward a desired direction, Kansei engineering cares for customer satisfaction that is both unbiased and scientific. A Kansei-engineered product does not need to be justified through flamboyant media or artistic packaging. It can be felt and experienced by the end user as it portrays the feelings of the end user, not of the artists or designers.

The ProEmotion system aims to understand the mental and emotional preference and inclination of the mobile phone users and to use the principal attributes (e.g., length, width, thickness and mass) of a mobile phone to predict its gender inclination. The assumption is the mobile manufacturers, rightly so, have always had in their minds about what group of customers their products target at, be it for male, female, general public or any other specific groups of users. ProEmotion can inform gender inclination of a mobile phone with accuracy as high as 90 %. This is based on a set of 92 mobile phone samples from the five major mobile phone manufacturers, Nokia, Samsung, LG, Motorola and Sony Ericsson.

The study also confirmed that the principal parameters of a mobile phone are playing an important role in its Kansei aspects. This quantified knowledge of user reactions and their relationship to the physical parameters of a mobile phone design aids industrial developers to understand the needs of the targeted consumers in the early stage of the development. It would not be a good practice for a designer to solely go for fashion and functionality with no due consideration of principal parameters.

This research may have answered, in part, the two questions posed at the beginning. The reason why some of the principal design attributes of mobile phones contribute to their Kansei remains a much tricky question to be answered. The accuracy level of ProEmotion can be further improved through the following means. More mobile samples can be included in the study. It could also be implemented into common industrial CAD software such as an add-on feature which extracts and analyzes emotional features of CAD models in real time. Analysis of more complicated or variety of features would also provide further insights into the product under study and add to the strength of this comprehensive software.

ProEmotion could also expand its product range to beyond analyzing only cellular phones. They could be programmed to automatically adjust its internal parameters to achieve optimal results depending on the user input database of other products. By achieving this, it could become a generic tool for product developers to apply Kansei engineering in design works.

The concept of Kansei engineering is relatively new in terms of industry application. Further research and development is still required. However, some successful cases in the industry have proven that this concept is not an imagery process but an enabling tool that can be used by industry to gain benefits in terms of higher market growth and better customer acceptance of the products.

References

1. Nagamachi M (1989) Kansei engineering. Kaibundo Publishing, Tokyo
2. Nagamachi M (1991) An image technology expert system and its application to design consultation. Int J Hum-Comput Interact 3(3):267–279
3. Nagamachi M (1995a) Kansei engineering: a new ergonomic consumer-oriented technology for product development. Int J Ind Ergon 15(1):3–11
4. Barnes C, Lillford SP (2009) Decision support for the design of affective products. J Eng Des 20(5):477–492
5. Choi K, Jun C (2007) A systematic approach to the Kansei factors of tactile sense regarding the surface roughness. Appl Ergon 38(1):53–63
6. Fukushima K, Kawata H, Fujiwara Y, Genno H (1995) Human sensory perception oriented image processing in a color copy system. Int J Ind Ergon 15(1):63–74
7. McDonagh D, Bruseberg A, Haslam C (2002) Visual product evaluation: exploring users' emotional relationships with products. Appl Ergon 33(3):231–240
8. Nagamachi M (2002) Kansei engineering as a powerful consumer-oriented technology for product development. Appl Ergon 33(3):289–294
9. Hohenegger J, Bufardi A, Xirouchakis P (2007) A new concept of compatibility structure in new product development. Adv Eng Inform 21(1):101–116
10. Nagamachi M (1995) Introduction of Kansei Engineering. J Indus Ergon 15(1):13–24
11. Ishihara S, Ishihara K, Nagamachi M (1998) Kansei inference system and internet VR. Manuf Hybrid Automation-II. J Indus Ergon 15(1):63–74, pp 403–406

12. Koda Y, Kanaya I, Sato K (2007) Modeling real objects for Kansei-based shape retrieval. Int J Autom Comput 4(1):14–17
13. Lee JWT, Chan SCF, Yeung DS (1995) Modelling constraints in part structures. Comput Ind Eng 28(3):645–657
14. Leon N (2009) The future of computer-aided innovation. Comput Ind 60(8):539–550
15. Peak RS, Lubell J, Srinivasan V, Waterbury SC (2004) STEP, XML, and UML: complementary technologies. J Comput Inf Sci Eng 4(4):379–390
16. Stouffs R (2008) Constructing design representations using a sortal approach. Adv Eng Inform 22(1):71–89
17. Wilson JR (1999) Virtual environments applications and applied ergonomics. Appl Ergonomics 30:3–9
18. Zhai LY, Khoo LP, Zhong ZW (2009) A dominance-based rough set approach to Kansei Engineering in product development. Expert Syst Appl 36(1):393–402
19. Kuang J, Jiang P (2009) Product platform design for a product family based on Kansei engineering. J Eng Des 20(6):589–607
20. Lai HH, Lin YC, Yeh CH, Wei CH (2006) User-oriented design for the optimal combination on product design. Int J Prod Econ 100(2):253–267
21. Lai HH, Chang YM, Chang HC (2005) A robust design approach for enhancing the feeling quality of a product: a car profile case study. Int J Ind Ergon 35(5):445–460
22. Lin YC, Lai HH, Yeh CH (2007) Consumer-oriented product form design based on fuzzy logic: a case study of mobile phones. Int J Ind Ergon 37(6):531–543
23. Matsubara Y, Nagamachi M (1997) Hybrid Kansei Engineering System and design support. Int J Ind Ergon 19(2):81–92
24. Nagamachi M (2000) Application of Kansei engineering and concurrent engineering to a cosmetic product. In: Proceedings of the ERGON-AXIA–2000, Warsaw, Poland, May 2000
25. Nagamachi M, Nishino T (1999) HousMall: an application of Kansei engineering to house design consultation. In: Proceedings of the international conference on TQM and human factors, Linkoping, Sweden, pp 349–354
26. Ogawa T, Nagai Y, Ikeda M (2009) An ontological approach to designers' idea explanation style: towards supporting the sharing of Kansei-ideas in textile design. Adv Eng Inform 23(2):157–164
27. Roy R, Goatman M, Khangura K (2009) User-centric design and Kansei Engineering. CIRP J Manuf Sci Technol 1(3):172–178
28. Schutte S, Eklund J (2005) Design of rocker switches for work-vehicles: an application of Kansei engineering. Appl Ergon 36(5):557–567
29. Akioka S, Fukumori H, Muraoka Y (2009) Search in the mood: the information filter based on ambiguous queries. Int J Comput Appl Technol 34(4):322–329
30. Bahn S, Lee C, Nam CS, Yun MH (2009) Incorporating affective customer needs for luxuriousness into product design attributes. Hum Factors Ergon Manuf 19(2):105–127

Chapter 8
Kansei Engineering: Methodology to the Project Oriented for the Customers

Viviane Gaspar Ribas El Marghani, Felipe Claus da Silva, Liriane Knapik and Marcos Augusto Verri

Abstract This work presents an original contribution to discussion on the theme Kansei Engineering (KE). This chapter intends to make a characterization of the theme KE and through it provide an understanding of the breadth of the subject. Beginning with what the literature shows about the subject and then focuses on which types of KE exist to be exploited, the survey sought information from the major bases of research papers available. The bibliographic is based on sources and database platforms available in academic pursuit (CAPES Periodicals) on the subject. Despite the rather ambitious goal, the aim is to bring the topic up for discussion, so that academics and practitioners together can reflect on this important theme, find better alternatives for interaction and look for appropriate ways of spreading this topic in Brazil. The research result is presented below, the text is structured so the initial highlight the context of KE, their definitions and basic guidelines and the existing types of KE and finally, a discussion.

Keywords Kansei Engineering • Method of design • Customer in process

8.1 Introduction

With the improvement of communication processes, consumers have become more demanding and the industry more competitive, so that only the functionality no longer meets the requirements of the market. Therefore, the current design

V. G. R. EL Marghani (✉)
Professor of Design of Universidade Federal do Paraná (Dr. Engineering), Paraná. Rua General Carneiro, 460 8° Andar, Curitiba, PR, Brazil
e-mail: viviane.gasparibas@ufpr.br
URL: http://www.design.ufpr.br

M. A. Verri
Student of Master in Design of Universidade Federal do Paraná (Dr. Engineering), Paraná. Rua General Carneiro, 460 8° Andar, Curitiba, PR, Brazil

F. C. da Silva · L. Knapik
Student of Graduate Design of Universidade Federal do Paraná (Dr. Engineering), Paraná. Rua General Carneiro, 460 8° Andar, Curitiba, PR, Brazil

S. Fukuda (ed.), *Emotional Engineering vol. 2*, DOI: 10.1007/978-1-4471-4984-2_8,   107
© Springer-Verlag London 2013

methodologies in the design field are looking for new methods allowing the generation of innovative ideas.

In this sense, the Kansei Engineering (KE) is presented as a methodology for product development able to translate the impressions, feelings and demands of users in solutions and parameters of concrete project to Schütte [1].

According to Matsubara and Nagamachi [2], KE is an effective technique to translate the feelings and desires of the consumer in product design elements. The consumer would be happier if this sensation experienced in the project could be implemented as Nagamachi [3].

In the future, it will become necessary to understand the user and the principles of how people interact with the world. Thus, professionals must acquire knowledge through user's experience and treat it as part of the system to optimize it.

To Schütte and Eklund [4], "Design and development of new products and product concepts has always been a challenge for companies on their markets. Internationalization, technological and economical development pressure contribute to an increasing competition in practice in all international markets. An increased number of products available, sometimes in combination with decreased purchasing power of the customers, forces companies to re-consider their product development strategies. Many examples support that their products change shape and connected to become closer to intangibles."

Although the KE has long exist, about 40 years, little is known about the methodology in the teaching project. In addition, the publications of KE in Brazil are rarely seen, and what is found is greatly simplified in terms of detail, compared with literature in other countries, where KE is detailed and applied broadly.

Thus, this work presents an initial discussion as a contribution by characterizing the KE topic and through it provides an understanding of the breadth of the subject.

It also encourages students to reflect on this important methodology so that together they can find alternative ways of implementing and diffusion of entrepreneurship in the country. As result, this work presents the context of the KE, their definitions and guidelines in addition to the basic differentiation between existing types identified. To end, it develops a discussion by the authors.

The novelty of this work will be highlighted insofar the result of the study approximates the method Kansei for the process of product design and fills a gap in domain of subjective requirements of users incorporated into the project. Moreover, as the applicability of other knowledge, approaches or methods— such as Ergonomics, Usability, Accessibility, and Quality Methods—the KE seeks to improve human relations with the objects in order to improve health, safety, effectiveness, efficiency, satisfaction, access, quality when products are used.

The incorporation of the Kansei method during the design process will also improve human relations with the objects, and will assist the production of products with higher added value, not just technicians, functional, esthetic and cultural, but also that satisfy the subjective desires and human aspirations.

8.2 Design and Emotion (Kansei)

The designer is a professional who intends to solve problems originated from the interfaces between man and his surroundings. He must develop all kinds of abilities and competencies to its maximum, acquiring knowledge from all areas, so that when he faces these problems, he will be able to distinguish the whole issue and its parts. He should also be willing to develop the capacity of "learn to see" noting the information emitted by all kinds of human expression with accuracy, in order to transfer these acquired experiences to his new conceived objects.

Nowadays, creativity is considered the heart of Design, consisting of the process of products' development in its design stages as a main ingredient; hence, it should be stimulated. Allied to creativity, the emotion component must be better understood, explored and applied in Design.

So, even if the utilization of Design Methods in Engineering provides satisfactory results to the designer, they are absolutely insufficient nowadays, because the consumer seeks, more and more, products that are *able to play the strings of the heart, stroke it's emotions, captivate culture and reason (...) able yet, to awaken it's active participation and at the same time, touch it*—Salocchi [5].

This causes the designer not only to develop his abilities, competencies and creativity, but also, above all, to mold them in the vision of the Italian designer [6], to whom *the object is a friend of men, a smiling and affectionate company.*

Moreover, in order for a professional of design to achieve outstanding projects, it is indispensable that he should be able to aggregate more value to his objects, not only the esthetic, technical, functional and cultural ones, but, mainly, the emotional ones. Thus, through the use of elements, like color, shape, texture, smells, flavors, sounds and movements; the object will be able to transmit to the consumer a vast amount of information.

To assure that consumer notices the implicit and explicit qualities of an object and to use this easier and more efficiently, it is necessary that he feels attracted and infatuated by it. Therefore, aggregating values to objects means to stimulate all areas that comprehend the human emotion: vision, hearing, tact, taste and smell; thus, the more sensations stimulated and awakened, more efficient the objects will be.

Norman [7], in his book *Emotion Design*, teaches *why beautiful objects really work better.* He gives us, among several examples, a teapot that has a whistle in its beak, which makes harmonious tune in the passage of steam. The consumer receives information that warns him that the tea is ready, and has his heart touched by a sweet melody.

A better understanding of the functioning of the human mind, as well as the brain information's processing according to Pinker [8] is needed in order to obtain results never reached before. The designer should gather enough knowledge to stimulate the human senses and, consequently, human emotion, not only his, as well as the consumer's.

Ramachandran and Hubbard [9] profited by the synesthesia, phenomenon that occurs when a person is able to mix the senses—vision, hearing, tact, taste and smell; being able, for example, to taste an image or to color a song—to incorporate it into the creative process of Design. The information given by objects will be more rapidly noticed due to the fact that more than one sense is being enhanced. As the consumer faces an object molded with feelings, he will understand the characteristics of this object, since his stimulated senses will awaken his soul and touch his heart. Unavoidably, the user will be amazed and will wish to possess this object.

As a creator, the designer is influenced by similar situations to those lived by the users of its objects. Which human being, when hearing the chord of a piano, or the sound of sea waves, or smelling a sweet perfume, or feeling solar rays, or even tasting a yummy strawberry pie covered with whip cream, has not created imaginary sets or mentally projected a new way of life?

8.3 Concepts and Definitions of Kansei Engineering

Since only the functionality no longer meets the requirements of the market, there is a need of the Design takes ownership of methods that allow the generation of innovative ideas to offer a differential to the industry, in accordance with Marghani et al. [10].

To Schütte and Eklund [4], the Japanese expression Kansei is difficult to translate. It means approximately "total emotions," but that does not fully explain its meaning—in fact not more than partly. Kansei is the impression somebody gets from a certain artifact, environment or situation using all their senses of sight, hearing, feeling, smell, taste as well as their recognition [3].

According to Schütte [1], the KE is a methodology that correlates the feelings and wishes of users with product development, and developing solutions and design parameters, so a consumer-oriented technology as [3], to incorporate solutions to the products. Still in accordance with Schütte [1], the KE is limited to the evaluation of words and their emotional impact on the human mind.

Ishihara et al. [11] define KE as a technology that seeks to translate human feelings into product features. For the authors, the Japanese word Kansei assumes the meaning of feeling, emotion and also printing. In the same year, Nagamachi [12] defines it as the transmission of feelings and image consumers have of a product design elements.

Ishihara et al. [11] also emphasize the importance of designers create products that meet the specific needs and feelings of users. Thinking that the user usually has the image or sense of how the product you want to buy, so it is important that designers identify them and turning them into products that meet the demand. And thus, the KE becomes a methodology in the intermediation between designers and consumers.

According to Nagamachi [12] and Nagamachi and Iamada [13], the KE was created in Japan, specifically the University of Hiroshima, in the 1970s. Japanese industries disseminated it by applying to the development of products, especially automobiles, homes, construction machinery, electrical appliances, clothes, household products, among others. The KE is defined as the translation of the "feeling" of the customer for the product (illustrated in Fig. 8.1).

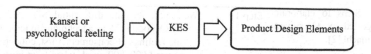

Fig. 8.1 Diagram of the process of Kansei Engineering system. *Source* Nagamachi [12]

Yet to Nagamachi [12] and Nagamachi and Iamada [13], KE includes the impressions of consumers about a product design, size, color, mechanical operation, ease of operation, as well as price. In 2001, Schütte and Eklund [14] treated the KE as a product development methodology that reflects the feelings, emotions and demands of customers for products and existing concepts in solutions and specific design parameters. This definition is similar to the previously mentioned authors.

Schütte and Eklund [14] define KE as a methodology that contributes to the systematic development of new and innovative products and can also be used as a methodology for improving product concepts. The KE is based on subjective estimates of properties and product concepts, and thereby helps consumers to express their needs, even though they have no conscience. Chuang and Ma [15] investigated how to express effectively in the process of developing new products, the image expected by users and got some questions.

- The designers really understand which are the expectations of users for the products?
- Users perceive this image that the designers tried to convey?
- There are formal characteristics that can be adopted to encourage the reporting of this expectation?

The use of the KE not only assists the designer in the creative process or in setting a desired image, but mainly in the perception gap between designers and users. With the use of tools and methods that bases the KE, we can develop a product with formal elements and characteristics that arouse the expected feelings by the users, increasing the chances of success of this product.

The translation of customer needs into design solutions and technical specifications is a major challenge to designers; the authors note [16]. Another challenge is to capture these emotional needs and their combination with design features that can meet them.

Hsu et al. [16] share the same opinion of Chuang and Ma [15], the question about if the designers understand the needs of the target audience, and if they are able to pass these expectations on the products they design, if the users/consumers perceive the effort of the designers and if not, how to express effectively to users/consumers the images they expect. In his work, Matsubara and Nagamachi [2] present steps and goals for the KE according to how he conceived it. The methodology provides a new product based on the feeling and the need of the consumer and must follow four basic steps (see Table 8.1), also defended by Schütte and Eklund [4].

Table 1 Steps to elaborate KE suggested by Matsubara and Ngamachi [2]	Step 1	We use the semantic differential (SD), developed by Osgood [17], as a technique to understand the feelings of consumers. First, collecting around 600–800 words taken from journals and interviews with professionals. They are selected from approximately 100 words most relevant.
	Step 2	A survey (or experiment) is conducted to find relationships between Kansei words and design elements.
	Step 3	With the aid of the computer is developed to a structure KES. Artificial intelligence, neural network and genetic algorithm as well as fuzzy logic are used to build a database and a computerized system of inference.
	Step 4	You can adjust the database Kansei words to the feelings of consumers to periodically update the data every 3 or 4 years.

1. Collect appropriate Kansei words to the product scope;
2. Identify the correlation between Kansei words and design features;
3. Building a technical identification of the design features;
4. Building a system that connects (1)–(3).

To Schütte and Eklund [4] also presents the three main points of KE are as follows:

1. Clarify the understanding of the consumer;
2. How to reflect and translate this understanding into product design?
3. How to create a system and organization-oriented to design Kansei?

Llinares and Page [18] define KE as a methodology for developing consumer-oriented products, which analyzes the characteristics of the product from the standpoint of the user. To them, goal is to identify and quantify the user perception about a product in their own configuration and find quantitative relations between these subjective responses and design features. The KE relies on the fact that the perception of users with regard to the product depends on the physical and psychological and therefore, the global evaluation given by a combination of both.

To Schütte and Eklund [14], the basic idea is to describe the concept behind a product from two different perspectives:

1. The semantic description;
2. The description of (product) properties.

Figure 8.2 shows the general procedures of the KE for Schütte [1].

To Schütte and Eklund [4], "these two descriptions span a vector space each. Subsequently these spaces are merged with each other in the synthesis phase indicating which of the product properties that evokes the different semantic impacts. First after these steps have been carried out, is it possible to conduct a validity test, including several types of Post-Hoc analyses. As a result from this step the two vector spaces are updated and the synthesis step is run again. When the results from this interaction

Fig. 8.2   Kansei Engineering procedure. *Source* Schütte [1]

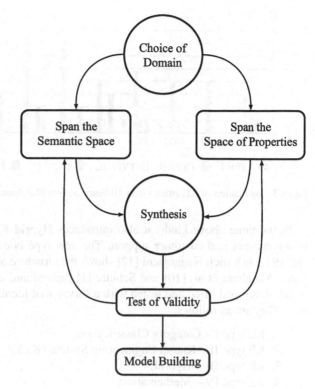

process are acceptable, a prediction model can be built describing how the semantic and the space of application are related."

8.4 Types Kansei

Statistical data submitted by KEB[1] in Fig. 8.3 show the growing publication on the subject enter 2001–2010, highlighting the importance of this methodology in the design process as it assists the development of products closer to consumers' expectations. The publications were divided according to the six existing types of KE, which will be posted throughout this topic.

In 1995, Nagamachi [12] classified KE into three types:

1. Category classification;
2. Kansei Engineering Computer System;
3. Kansei Engineering Modeling.

---

[1] O *KEB group* é constituído por professores e estudantes da área de design da instituição brasileiras de ensino UFPR. Tem realizado levantamentos e discussões sobre o tema, os quais evidenciam a importância da utilização do KE como suporte para a tomada de decisões no processo de projeto, ou seja, no desenvolvimento de produtos.

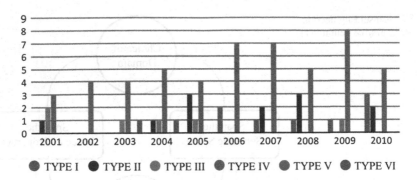

Fig. 8.3 Publication of KE articles over 10 Years. *Source* Marghani et al. [10]

In the three above kinds, it also introduces Hybrid KE, a combination of KE as a consumer and customer support. The new type is exemplified in a later article, 1997, in which Nagamachi [12] shows the structure along with Matsubara [2]. Later, Marghani et al. [10] and Schütte [1] present and characterize the six types of KE developed, tried and tested with a survey that identified the published literature. They are as follows:

1. KE type I—Category Classification;
2. KE type II—Kansei Engineering System (KES);
3. KE type III—Hybrid;
4. KE type IV—Mathematics;
5. KE type V—Virtual;
6. KE type VI—Collaborative.

8.4.1 KE Type I: Category Classification

According to Nagamachi [12], the KE type I is a method in which a Kansei category of products is divided into a tree structure to identify the design details. This division should result in sub-concepts with clear meanings to achieve the required identification. It is a fast and easy way to analyze KE, as says [1]. In this type, a particular product strategy and market segment are identified and developed in a tree structure (see Fig. 8.4). This structure presents similarities with Ishikawa's fishbone diagram [19] and QFD [20].

Through the Kansei words, consumers are encouraged to express their feelings and their emotional states. A method of analysis used is differential semantic (DS).

8.4.2 KE Type II: Kansei Engineering System

In KE type II, feelings and property of users of the products are stored in a digital database. Then, it made the correlation with of the feelings with pictures of the

Fig. 8.4 KE process tree. *Source* adapted from Nagamachi [12] and Schütte [1]

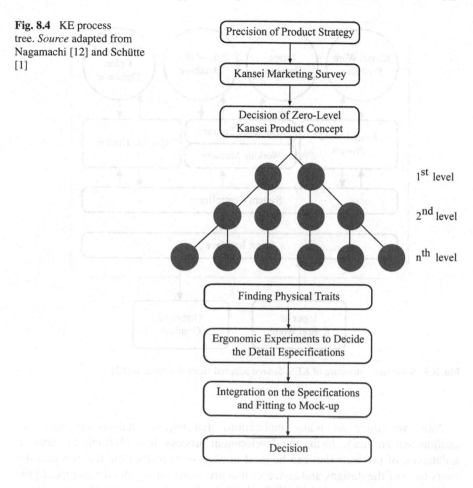

products analyzed. The KES are, in many cases, specialized programs to support the design decisions of different products.

According to Schütte [1], a typical KES consists of four databases: Kansei words, images of the analyzed products, design and color as well as knowledge about how the different data are related. Its basic structure is illustrated in Fig. 8.5.

8.4.3 KE Type III: Hybrid

The KE type III is called hybrid because it is constituted by KES and reverse Kansei Engineering. According to Schütte [1], KES is also known as Forward Kansei Engineering, because it can only be used to convert consumer feelings into design parameters.

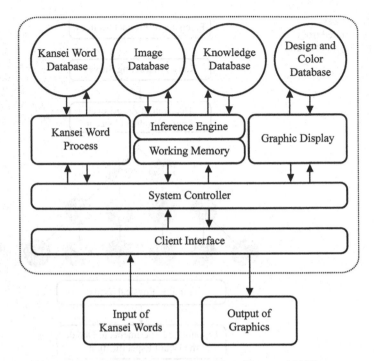

Fig. 8.5 Systematic structure of KES. *Source* adapted from Nagamachi [12]

Moreover, there are many applications that require analysis concepts and designs that are made during the development process. It is advisable to maintain a database of products that can be used in reverse to predict the feelings that the users have of the designs and concepts that are made during the development process. Thus, this type is called KE Backward Kansei Engineering System, whose structure can be seen in Fig. 8.6.

Although the databases used for the Hybrid KE be the same as the KES, the program is specifically designed for use by designers that feed the system with their ideas through the user interface and system controller, which in turn examines the parameters of the product and compares them with the data stored in the database. These data are related to the database Kansei words that are selected and presented to the designer in order to help generate new ideas, says [10].

Schütte [1] presents several advantages such as the Hybrid KE: rapid assessment of clients Kansei on the concept, it is not necessary to present the concepts or prototypes for potential users as it is not necessary an expensive market research.

Marghani et al. [10] warn off the complexity of this type of KE, since a number of other functions are implemented. It is possible, for example, include a system for recognition of shape and color in order to analyze the characteristics of design drawings of the new product.

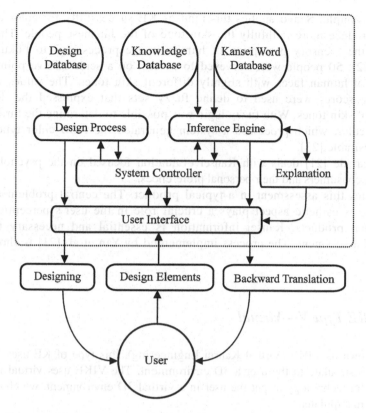

Fig. 8.6 Structure of Backward KES. *Source* adapted from Nagamachi [3]

8.4.4 KE Type IV: Mathematical

The preferences of consumers as stipulated by the Kansei attributes evaluated in a product vary according to each person, according to Yan et al. [21]. For the authors, the evaluation Kansei of consumers is vague due to the different priorities that each respondent gives to every word Kansei, which ultimately generates an uncertainty on the data obtained.

In accordance with Schütte [1], the differential of the type IV KE is the application of a mathematical model. Marghani et al. [10] indicate a strong trend in the use of this type of KE as it allows translation of inaccurate data from users on accurate data. Also according to Marghani et al. [10], this feature actually helps the project team to have more precise conditions for making decisions regarding the development of the project.

The KE type IV uses more than intelligent systems; in particular, this type uses a mathematical model. This model is built from the rule base to obtain the result of ergonomic Kansei words. In this procedure, a mathematical model implies a kind of logic that plays a role similar to the rule base according to Nagamachi [12].

For example, Sanyo applies this kind of KE successfully for color copies that reproduce more faithfully the skin tone of the Japanese people. The first step is the "sensory perception of human image processing" to Fukushima et al. [22]. 50 people were surveyed to assess, on a scale of five points, 24 images of human faces with slightly different skin tones. The values of the resulting scores were used to define fuzzy sets that expressed the desired degree of skin tones. With fuzzy logic was possible to determine the three factors of color, which processed a system generated the RGB color type skin more desirable [23].

The article [21] deals with Kansei evaluation focused on the psychological needs of consumers and their personal preferences.

Applies this assessment in a typical product. The central problem of the work is: as esthetic aspect plays a crucial role in the user's perceptions of traditional products, Kansei information is essential and necessary to the issue of assessment. The process implemented by Yan et al. [21] is shown in Fig. 8.7.

8.4.5 KE Type V—Virtual

Also known as VIKE (Virtual Kansei Engineering), this type of KE uses virtual reality to simulate to the user a 3D environment. The VIKE uses virtual reality, a powerful technology to put the user in a virtual 3D environment, which can be directly manipulated.

It is a combination of a computer system with virtual reality systems to help the user's selection of a product using his experience as a resource to the virtual space [23]. According to Nagamachi [3], by using this type of KE, it is possible to check the Kansei environment before the production of the Kansei product which makes this technology a combination very useful and effective for the design of houses, car interiors and urban design and landscape.

It has been as an example the application of this methodology in the development of kitchens and dining rooms by the Matsushita Company in partnership with the University of Hiroshima [24]. The system on the left represents Kansei Engineering, and the system on the right, the virtual system (see Fig. 8.8).

First, the consumer answers to the questions regarding their lifestyle and inserts his words Kansei. The system proposes the kitchen that fits the user's Kansei and used a database based on the feelings generated by the project kitchen imagined by 10,000 women. If the consumer is satisfied in the virtual space, the designed kitchen decided by the system is transferred to the factory and delivered in two weeks. The new kitchen is mounted in the presence of the consumer. This system is very popular in Tokyo, Nagoya, Osaka and Hiroshima according to Nagamachi [3].

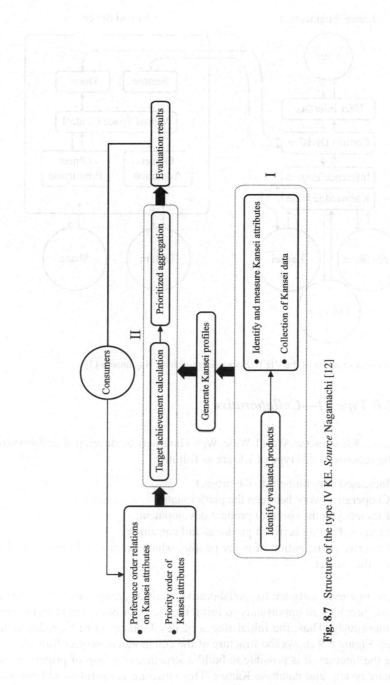

Fig. 8.7 Structure of the type IV KE. *Source* Nagamachi [12]

Kansei Engineering   Virtual Space

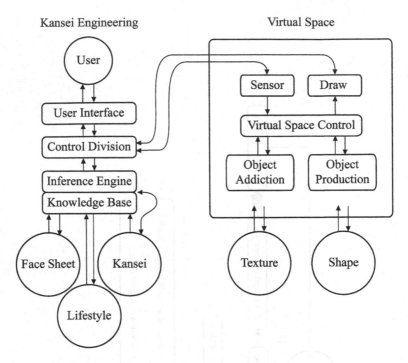

Fig. 8.8 System structure of the VIKE. *Source* adapted from Nagamachi [3]

8.4.6 KE Type VI—Collaborative

This type of KE uses the World Wide Web (Internet) to develop a collaborative work. The benefits of this type of KE are as follows:

- Increased commitment to the project;
- Cooperative work between the participants;
- Efficiency in the speed of product development;
- Increased dialog between producer and consumer;
- Effective participation of many people, which promotes diversity of ideas for the project.

It is a type that uses software for collaborative work in groups or on the Internet. In this case provides an opportunity to bring the views of both clients and designers simultaneously. Thus, the initial stages of development must be reduced and simplified. Figure 8.9 shows the structure of the collaborative project Kansei.

Using the Internet, it is possible to build a structure of group of project, which has a smart system and database Kansei. This structure is useful to add the work of many designers. The system, through a server, is a clever program that supports collaborative project allowing greater interaction between designers and project development [3]. As said before, the benefits of this type of KE are, for example,

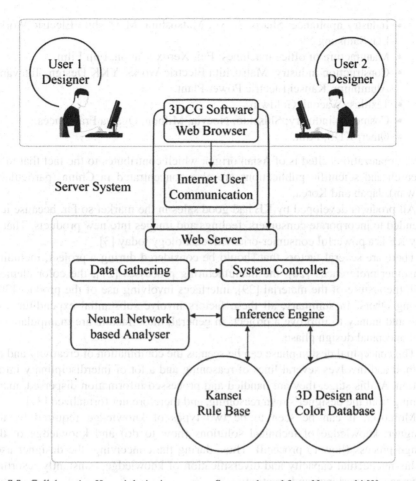

Fig. 8.9   Collaborative Kansei designing system. *Source* adapted from Nagamachi [3]

collaborative work among participants, efficiency, speed product development, greater dialog between producer and consumer, and the effective participation of many people, offering a diversity of ideas for the project, as says [25].

8.5 Considerations

Many products are developed in Japan using Kansei Engineering. Recently, it was applied to construction products, as well as urban projects. These are some organizations, not only Japanese, which introduced Kansei Engineering by the year 2002, according to Nagamachi [3]:

- Automotive: Mitsubishi, Mazda, Toyota, Honda, Ford, Hyundai, Delphi Automotive Systems.
- Industry of construction machines: Komatsu.

- Industry appliance: Sharp, Sanyo, Matsushita, Matsushita Electric Works, LG, Samsung.
- Manufacture of office machines: Fuji Xerox, Canon, Fuji Film.
- Construction industry: Matsushita Electric Works, YKK Design, Tateyama Aluminum, Kansei Electric Power Plant.
- Textiles: Wacoal, Goldwin.
- Cosmetics industry: Shiseido, Noevia, Milbon, Ogawa Fragrances.
- Other: KDS, Pilot.

Most organizations cited is of Asian origin, which contributes to the fact that most research and scientific publications are also concentrated in China (particularly Taiwan), Japan and Korea.

All products developed by KE had good sales in the market so far, because it is intended to incorporate consumers' feelings and images into new products. That is why KE is a powerful consumer-oriented technology today [3].

There are several factors that should be considered during a project, including consumer preferences [26], the manufacturing processes [27], the color element [28], the texture of the material [29], interfaces involving use of the product [30], among others. In summary, all these factors involve substantial expenditures of time and money to improve a project; in general, these factors are manipulated in the conceptual design phase.

The conceptual design phase can be seen as the combination of creativity and no method and involves several lines of reasoning and a lot of interdisciplinary information. At this stage, they are handled and processed information dispersed, many coming from the designer's own reasoning, and therefore not formalized [31].

Moreover, it can be seen to be two types of knowledge required by the designer: knowledge of technical solutions (how to do) and knowledge of the design process (how to proceed). Thus, during the conceiving, the designer uses all his intellectual capacity and diversification of knowledge, constantly resorting to personal experience and creativity.

Thus, the KE is a positive methodology, it corroborates with the decisions of the project team that uses more precise information from the data manipulation by users who have departed. With this design process is increased efficiency not only in its conceptual stage, but the production phase, since the products are in best accordance with the needs of customers/users. The methodology KE helps the theories of decision making, as it enables the project team visualize more accurately the needs and desires of customers/users, so the project becomes faster, accurate and results in next human aspirations.

In the Design, the subjectivity is an area not totally dominated, or the understanding of perception that customers/users give to objects [10]. The use of the KE not only assists the designer in the creative process or in setting a desired image, but mainly in reducing the perception gap between designers and users. With the use of tools and methods that bases the KE, we develop a product with formal elements and characteristics that arouse the expected feelings in the users, increasing the chances of success of this product.

Despite the large international efforts targeting these areas, which reinforce its importance as a field of current research, it appears still underdeveloped, without knowledge totally sedimented. Because it is a strategic area in the field of design knowledge, and rapidly developing, its importance should have been emphasized and cannot dispense with investments in any company that is interested in participating in the international community as the owner of knowledge and innovation.

In Brazil, little or nothing about this subject is addressed. Thus, the Brazilian situation presents itself lagging behind the more developed centers internationally. However, because it is still a developing knowledge, this situation can and must be changed.

It was intended in this article to demonstrate that the understanding of human's feelings and emotions must have its importance enhanced by the designer during his academic and professional training, which may be an essential ingredient in developing new projects. When faced with the challenges presented in a briefing, the professional must master the skills that will facilitate the generation of innovative ideas. And this is not only in terms of functional and esthetic techniques, but above all emotional.

To this end, it was demonstrated that the KE can collaborate on creating repertoires of knowledge by using this methodology. As the KE can help the designer to highlight the qualities of an object and on the development of new products, this methodology must be considered and studied. Thus, the difference in product, even though technically equates to that of its competitor, lies in the professionalism with which the design team handles all the details of a project, such as the pianist gives himself to music.

The designer you want to succeed, it is essential to learn to look around, listen to the best colors, shapes feel, taste sounds, experimenting with various aspects of the world. Due mainly to "wake up" for the discovery and formulation of solutions to the problems presented.

The designer must also understand all the ways of life and human expression, be it musical, literary, poetic, sensitive and emotional to convey ideas, feelings about the world, human desires, and more than that, try to help each other solve the problems of human and environmental diversity.

And finally, being a designer is a practice of life, a privileged way of linking the intelligence, the thought of leading, guiding the formation of the culture of human beings, the real, sensible and inimaginary because through their methods of solving problems are distinguished divergent and convergent thinking, logic and structure, mobilizing structures and moving in search of new, or worth the KE can be produced industrially and operationally to build a decent and comfortable place for humanity.

References

1. Schütte S (2002) Designing feelings into products. Integrating Kansei engineering methodology in product development. Linköping: Linköping studies in science and technology. Thesis No. 946, ISBN: 91-7373-347-4 ISSN: 0280–7971
2. Matsubara Y, Nagamachi M (1997) Hybrid Kansei engineering system and design support. Int J Ind Ergon 19:81–92

3. Nagamachi M (2002) Kansei Engineering as a powerful consumer-oriented technology for product development. Appl Ergon 33:289–294

4. Schütte S, Eklund J (2003) Product development for heart and soul. Linköpings Universitet, Department for Human Systems Engineering, Sweden Press, Uni Tryck Linköping/Sweden ISBN: 91-631-4295-3

5. Salocchi C (2004) Gemma Gioielli [Online] Available from URL: <http://www.gemmagioielli.com/inglese/index.htm. >. [Accessed 2004 May 26th]

6. Mendini A (2004) [Online] Available from URL: <http://www.the-artists.org/ArtistView.cfm ?id=8A01F9C6-BBCF-11D4-A93500D0B7069B40>. [Accessed 2004 May 26th]

7. Norman DA (2008) Design emocional. Por que adoramos (ou detestamos) os objetos do dia-a-dia. Rio de Janeiro: Rocco

8. Pinker S (1999) How the mind works. W.W. Norton & Company, New York. ISBN: 0393318486

9. Ramachandran VS, Hubbard EM (2003) Ouvindo as cores e degustando as formas. Brasil, São Paulo, Scientific American, ano 02, vol 13, pp. 49–55

10. Marghani VGREL et al. (2011) Kansei Engineering: metodologia orientada ao consumidor para suporte a decisão de projeto. Congresso Brasileiro de Gestão e Desenvolvimento de Produto

11. Ishihara HH et al (1995) An automatic builder for a Kansei Engineering expert system using self-organizing neural networks. Int J Ind Ergon 15:13–24

12. Nagamachi M (1995) Kansei Engineering: a new ergonomic consumer-oriented technology for product development. Int J Ind Ergon 15:3–11

13. Nagamachi M, Iamada AS (1995) Kansei Engineering: an ergonomic technology for product development. Int J Ind Ergon 15:1

14. Schütte S, Eklund J (2001) An approach to Kansei Engineering—methods and a case of study on design identity. Asean Academic Press, London

15. Chuang MC, Ma YC (2001) Expressing the expected product images in product design of micro-electronic products. Int J Ind Ergon 27:233–245

16. Hsu SH, Chuang MC, Chang CC (2000) A semantic differential study of designer's and user's product form perception. Int J Ind Ergon 25:375–391

17. Osgood CE (1969) The nature and measurement of meaning. In: Osgood CE, Snider JG (eds) Semantic differential technique—a source book. Aldine publishing company, Chicago, pp 3–41

18. Llinares C, Page A (2007) Application of product differential semantics to quantify purchaser perceptions in housing assessment. Building Environ 2011(IC 736307):13 doi:10.1155/2011/736307

19. Ishikawa K (1982) Guide to quality control. Asian Productivity Organization, Tokyo

20. Akao Y (1996) Introdução ao desdobramento da qualidade. Belo Horizonte: UFMG, Escola de Engenharia, Fundação Christiano Ottoni

21. Yan HB et al (2008) Kansei evaluation based on prioritized multi-attribute fuzzy target-oriented decision analysis. Inf Sci 178:4080–4093

22. Fukushima K, Kawata Y, Genno H (1995) Human sensory perception oriented image processing in a color copy system. Int J Ind Ergon 15:63–74

23. Nagamachi M (1996) Kansei Engineering and implementation of human-oriented product design. In: Koubek RJ, Karwowski W (eds) Manufacturing Agility and Hybrid Automation. IEA Press, USA, pp 77–80

24. Enomoto N et al (1995) Kitchen planning system using Kansei VR. In: Symbiosis of Human and Artifact, pp 185–190

25. Nagamachi M, Nishino T (1999) HousMall: an application of kansei engineering to house design consultation. In: Proceedings of the international conference of TQM and human factors, pp 349–354

26. Hong SW, Han SH, Kim KJ (2008) Optimal balancing of multiple affective satisfaction dimensions: a case study on mobile phones. Int J Ind Ergon 38:272–279

27. Pine BJ (1993) Mass customization: the new Frontier in business competition. Harvard Business School Press, Boston
28. Hsiao SW, Chiu FY, Chen CS (2008) Applying aesthetics measurement to product design. Int J Ind Ergon 38:910–920
29. Chang CC (2008) Factors influencing visual comfort appreciation of the product form of digital cameras. Int J Ind Ergon 38:1007–1016
30. Artacho-Ramírez MA, Diego-Mas JA, Alcaide-Marzal J (2008) Influence of the mode of graphical design representation on the perception of product aesthetic and emotional features: an exploratory study. Int J Ind Ergon 38:942–952
31. Durkin J, Durkin J (1998) Expert systems—design and development. Prentice Hall, New York

Chapter 9
Kansei Engineering: Types of this Methodology

Viviane Gaspar Ribas EL Marghani, Felipe Claus da Silva, Liriane Knapik and Marcos Augusto Verri

Abstract This chapter presents the six (6) types of Kansei engineering (KE) in more details than in chapter Kansei engineering: methodology to the project oriented to the consumers. Studies of the application of each type will be presented, since its inception to 2011. Thus, it is possible to notice the different procedures that each type can assume; in addition, the answers may be obtained by applying this methodology. Each type of KE uses different supports to its procedures, for example, mathematical models, virtual simulation of product usage or environment, and the joint work of designers supported by a server with intelligent software. Throughout this chapter, it is possible to realize that the application of KE includes since the macro until the micro-products/environments, the KE can be applied on the users' preferences about a flat, the interior of a car or small objects, like a cell phone. At last, that chapter is concluded with a discussion concerning the importance of KE to business and industry, what justifies its discussion and dissemination.

Keywords Kansei engineering • Method of design • Customer in process

9.1 Introdution

All the six (06) types of Kansei engineering (KE) developed today have different procedures and rely on the use of appropriate methods for each one. Thus, the KE advances are directed to detailing responses obtained by its application.

According to Schütte [1], the KE can be carried out in different ways using different types of Kansei engineering. At the moment, six (06) types of KE have been developed, tested and proved, and they are as follows:

V. G. R. EL Marghani (✉)
Dra. Engineering, Federal University of Paraná, Rua General Carneiro 460,8° andar,
Curitiba, Paraná, Brazil
e-mail: viviane.gasparibas@ufpr.br
URL: http://www.design.ufpr.br

F. C. da Silva · L. Knapik · M. A. Verri
Federal University of Paraná, Curitiba, Paraná, Brazil

S. Fukuda (ed.), *Emotional Engineering vol. 2*, DOI: 10.1007/978-1-4471-4984-2_9,   127
© Springer-Verlag London 2013

1. KE type I—Category classification;
2. KE type II—Kansei engineering system (KES);
3. KE type III—Hybrid;
4. KE type IV—Mathematics;
5. KE type V—Virtual;
6. KE type VI—Collaborative.

The type of KE should be selected according to the desired outputs, such as: What is the genetic code optimum of a product? The time available that the project team can devote to the development of procedures, analysis and reflection on the results should also guide the choice of the type.

Besides these factors, it must be also consider the time for training and practice for the type chosen to develop a product, or just to analyze the human feeling of a certain product. These variables are crucial for any project team wishing to use the methodology KE.

9.2 KE Type I

According to the author Schütte [1], in the KE type I, a product strategy and market segment are identified and developed in a tree structure. The application of this type begins with the concept known as zero (0), which will be divided into several subconcepts at the following levels. These concepts can be evaluated separately in different levels until the design parameters of the product can be easily determined.

In more complex cases, it may be necessary to carry out surveys and/or researches that support the decision for a certain concept of the final product. The tree structure can be seen in Fig. 9.1.

The definition of KE type I mentioned above, according to Schütte [1], can be perceived both in the work done by Nagamachi [2] and in the work of the authors Jindo and Hirasago [3] in 1997, in which the KE type I was applied. These authors had aimed to understand the uncertain needs of users and develop the products, in both cases were cars, based on the Kansei words selected by each user.

The work was based on data relating to the design of car interiors, the design elements, and the human impressions conveyed by the elements of design. Figure 9.2 illustrates a flow diagram of the construction and a support structure for the styling of a car.

The author Schütte [1] reports that the KE type I depends, in many cases, not only on a property of the product, but also on the composition and the balance between properties of components of the product. Therefore, for the application of KE type I, is required the division of a product into its basic units (components and subcomponents), using or not a tree structure, so that the product can be analyzed in its entirety from its elements.

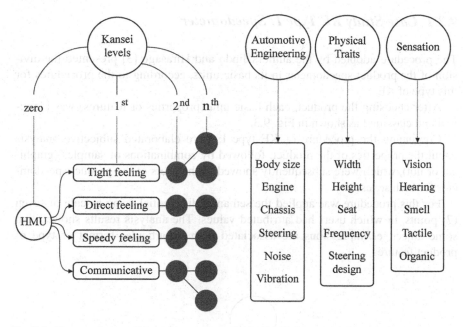

Fig. 9.1 The translation of Kansei into car physical traits in the case of 'Myata'. *Source* adapted from Nagamachi [2]

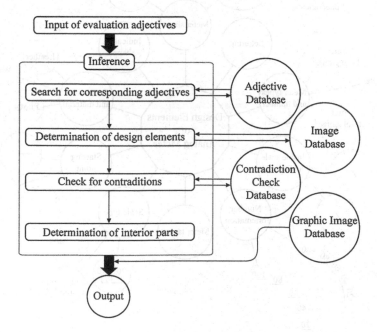

Fig. 9.2 Construction and flowchart of car interior styling support system. *Source* adapted from Jindo and Hirasago [3]

9.2.1 Case Study KE Type I: Speedometer

The procedure adopted by the authors Jindo and Hirasago [3] presented the division of the product speedometer in its basic units, according to the procedures for this type of KE.

After choosing the product, each basic unit (properties or features) was identified and classified as shown in Fig. 9.3.

Continuing the procedure of KE type I, were elaborated subjective analysis about the properties of the product, followed by combinations of samples (graphical or not), which were subsequently showed to the users of the product speedometer for both sexes.

For this procedure was applied the semantic differential using a scale of seven (7) points, in which users had attributed values. The analysis results showed that some adjectives/impressions are associated with a particular format or type of product feature.

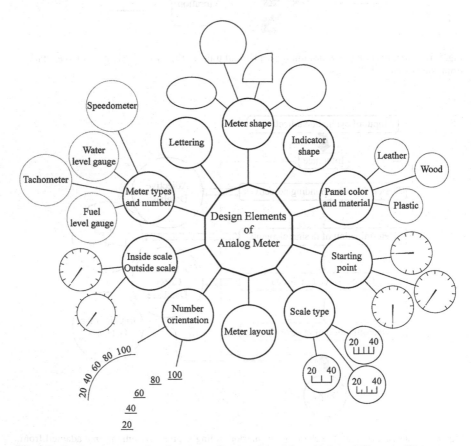

Fig. 9.3 Design elements of analog meter. *Source* adapted from Jindo and Hirasago [3]

Thus, it was identified the most desirable shapes to users, considering what their needs are. The result indicated that speedometers should adopt shapes that transmit to the users the impression of easy understanding, that is, the best design should transmit that feeling. It was obtained as result words/adjectives and forms/formats that represent that concept.

The number of words to be used in the semantic differential scale depends on the team and also on the project objectives, and depends on the product and on how many elements the product has, among other variables. According to the authors Chuang and Ma [4], the purpose is to apply the semantic differential scale to quantify the user perception. Thus, it is possible to relate the subjective responses with the design parameters.

At a given time was presented to the respondents photographs of speedometers. It was used the concept easy understanding as the axis, in which other words were situated.

The user based its assessment on the design criteria that would classify as easy to understand. The higher the number of respondents, the most consistent may be the results, according to the authors Chuang and Ma [4].

In the study presented by the authors Jindo and Hirasago [3], a vehicle characterized as comfortable, the speedometer must be easily understood, and the seat must be ergonomically designed, among other features.

Thus, the study indicates that different parts of the same product can have different impressions on the user, and the individual analysis of these units is needed to fully meet the desires of the user.

As a conclusion of the case analyzed, it can be said that the units of the product do not need to arouse the same feelings that product as a whole stimulates in the user, because each party meets its own characteristics. The subdivision of the product components in their minimum units specifies better the relevance, the needs and desires of the user to facilitate the improvements that are made in product design.

9.3 KE Type II: Engineering System (KES)

The development of KE type II implies the application of KE type I as an initial procedure. After the feelings of users were identified through the Kansei adjectives, such information is stored in digital databases to make correlations between Kansei words with images of the products that were analyzed.

This allows designers to know better the understanding of the perception that users have with respect to products and guides the designers to make decisions closer to what users want.

According to Nagamachi [2], these databases are separated into four (4), which are related categories as can be seen in Fig. 9.4.

For the development of the procedure KES, there are steps that take place as follows, according to Nagamachi [2]:

Fig. 9.4 Systematic
Structure of KES. *Source*
adapted from Nagamachi [2]

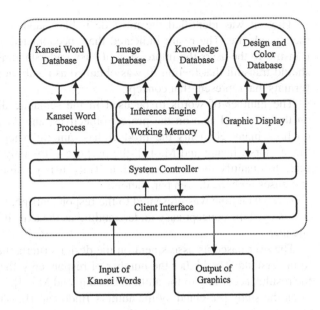

1. Insert the words in Kansei KES (the consumer should think of the desired product);
2. Check the Kansei word recognition with the database;
3. If the words were recognized, then they are transferred to the knowledge database;
4. Activate the inference engine, which will inform, on a screen, what aspects of design details and colors are appropriate, that is, are in accord with the feeling of the user.

9.3.1 Kansei Database

After collecting the Kansei words and the selection of these words, semantic differential scales are developed and applied. Later, the data can be assessed by factorial analysis or other methods of analysis. The results suggest Kansei words, giving rise to the database built into the system database.

9.3.1.1 Image Database

The results of the differential semantic can be analyzed by the quantitative theory type I [5], which is a type of multiple regression analysis for qualitative data.

With this analysis, it is possible to obtain a list of statistics that relate Kansei words and design elements (product properties). Thus, the elements that actually contribute to meeting a specific Kansei word can be identified in the products.

9.3.1.2 Database Design and Color

The details of the product properties that relate to the colors and design are placed in a database separately. The details of design consist on the formal aspects of the product (object of study) that can be correlated to a Kansei word. The details of colors correspond to the relations of Kansei words with colors that can be transmitted to users. All combinations of design details and colors are set by the system database, which allows the visualization of graphs for analysis.

9.3.1.3 Database Knowledge

In this database, the rules necessary to decide which items are correlated to the details of design and color and Kansei words are established. Some rules, for example, may result from the calculation of the quantity theory, or on some principles of colors, or rules about the forms of objects and their symbols.

9.3.2 Case Study: Urban Flats

In their work, the authors Llinares and Page [6] used the KE type II to analyze the users' perception in relation to urban flats (apartments) in Valencia, Spain. The authors applied the semantic differential and identified fifteen (15) major factors that characterized the perception of users. In this study involved one hundred fifty-five (155) subjects who analyzed one hundred twelve (112) flats.

This work demonstrates the breadth of application of the methodology KE, because it can assist both the analysis of product design and architectural projects.

One of the authors' conclusions is the assertion that the methodology KE helped them identify key concepts and attributes that describe the users' perception about a specific property, which are identified by the expression of words used by users.

To obtain users' data were used questionnaires to collect information about respondents and their perceptions. To identify the characteristics of the flat were collected sixty (60) words taken from magazines and journals on the subject as shown in Fig. 9.5.

For processing the data, the techniques used were as follows:

- Identification of a semantic axis;
- Classification of semantic axes;
- Comparative analyzes.

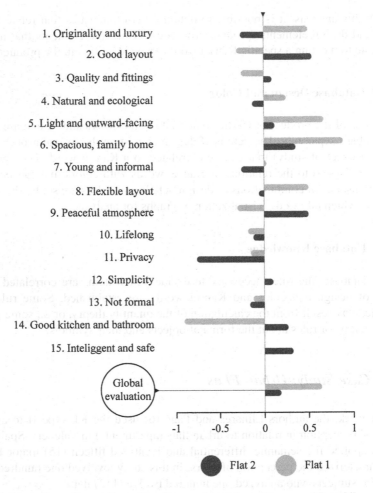

Fig. 9.5 Comparison of semantic profiles. *Source* adapted from Llinares and Page [6]

The collected data were then stored and grouped in databases, the main characteristic of the KE type II.

9.4 KE Type III: Hybrid

The KE type III is termed hybrid because it consists on the KES and reverse engineering. According to the author Schütte [1], KES is also known as forward Kansei engineering, because it can only be used to convert consumer sentiment in design parameters.

On the other hand, there are many applications requiring analysis of concepts and designs that may be developed during the process of product development. For this reason, it is advisable to maintain a database of products that can be used in reverse order.

This procedure allows to predict the feelings that the designs and concepts that were developed during the development process will awake on the users. Thus, it is possible to get a return from the user and verify whether the product actually meets the requirements of the project and the Kansei words defined by users.

To the authors Roy et al. [7], this process occurs by the development of seven (7) steps, as shown in Table 9.1.

The first step identifies the properties of the product to be designed based on a survey data from similar products. This research should select provided characteristics for the product and select the similar which best meets the similar goal of the project.

After these steps, there should be a selection of features that best meet the needs of the project.

In the second stage are selected Kansei words, which can be collected from various media such as Web sites, magazines or opinions of the users. After collecting the Kansei words, they must be refined from an affinity diagram.

Thus, the number of words is reduced, and the Kansei words with higher affinity will have greater relevance to the project.

In the third step is applied a questionnaire with the Kansei words selected on the first stage and the objects selected on second stage. Besides the subjects respond about the product, it is possible, with the use of Kansei words, ask them to describe the ideal product. In the case of this research, the product evaluated by Roy et al. [7] was a cell phone.

In the fourth stage, the results are analyzed to understand the perception of users and translate this information into design parameters. In the fifth stage, the alternatives are generated for the new product from the requirements of the fourth stage. The alternatives are selected among the project team based on product requirements.

Table 9.1 Etapas do processo para [7]		
	Step 1	Identify properties of products
	Step 2	Identify Kansei words
	Step 3	Match the words with the Kansei product in a questionnaire
	Step 4	Analyze the results of the questionnaire
	Step 5	Use the analysis for generating alternatives
	Step 6	Make a second questionnaire on the options selected using the same words of first questionnaire
	Step 7	Discuss the results and verify whether it meets the users' perceptions, otherwise return to first step

In the sixth step, another questionnaire is applied with the selected alternatives using the same words and Kansei of the third stage. Thus, it is possible to verify whether the design team was able to understand the needs and desires of users and if the alternatives reflect this expectation. Understanding the perceptions of users regarding the product is the focus of this type of KE, because it needs this understanding to transform all the sentiments expressed into a product.

In the seventh step, the design team analyzes the survey results more deeply and carefully, and then the product is created. Finally, the user is queried to determine whether the proposal meets your expectations. If the goal was reached, the production is allowed; otherwise, it must return to the first step.

9.4.1 Case Study: Mobile Phones

The authors, Roy et al. [7] applied the procedures of KE type III to cell phones. The first research was focused on three brands considered the best ones by users: Nokia, Samsung and Sony Ericsson. The procedures reported in the KE type III were applied, and the results obtained were converted into alternatives generated by the last questionnaires (see Fig. 9.6). The alternatives had successful in almost all requirements indicated by the users, except durability and clarity.

The questionnaire shown in Fig. 9.6 was first used in contact with the subjects through the use of similar products, and then a second contact was developed to compare results based on responses from both questionnaires. According to Roy et al. [7], when this step is finished and the expected results were not achieved, it is necessary to redo all the steps, find the critical points and make the necessary changes.

9.5 KE Type IV: Mathematical

Consumer preferences regarding the Kansei attributes evaluated in a product vary according to each person. According to the authors Yan et al. [8], Kansei evaluation of consumers is vague due to the different priorities that each respondent gives each Kansei word, which creates uncertainty about the data.

According to Schütte [1], the differential of the KE type IV is the implementation of a mathematical model. For this feature, the authors El Marghani et al. [9] indicate a strong trend in the use of this type of KE as it allows translation of inaccurate data on accurate data from users, that is, mathematical results.

The authors El Marghani et al [9] claim that this feature actually helps the project team in better conditions for making decisions about the development of a product.

Fig. 9.6 Example of questionnaire applied via Web

9.5.1 Case Study: Leaves of Gold Kanazawa

In this study, the authors Yan et al. [8] report the successful implementation of KE type IV in different types of consumer products, but also highlight the lack of attention given to its application in commercial products, especially when it comes to the traditional arts.

The study presented by the authors is the Kansei evaluation of thirty (30) Kanazawa gold leaf, a traditional art with more than 400 years of history. This assessment is justified by the contribution it brings to the marketing area, besides the fact that KE type IV is crucial for understanding the esthetic perception of consumers by allowing an accurate translation of the desires of the users in mathematical variables.

9.5.1.1 Methodology

The authors Yan et al. [8] propose a mathematical model for the KE, consisting of four (4) main steps:

1. Identification and measurement of Kansei attributes;
2. Develop Kansei profiles;
3. Specification of consumer preferences;
4. Aggregation prioritized fulfillment of goals.

Also according to the authors Yan et al. [8], this methodology allows consumers to select and purchase products according to their actual preference.

9.5.1.2 Identification and Measuring Attributes Kansei

To identify the Kansei attributes, we developed a brainstorming with manufacturers and local tenants and, as a result, obtained twenty-six (26) pairs of Kansei adjectives. Linguistic variables were used to express the semantic differential in a scale of seven (7) points, and each attribute Kansei was represented by a triangular fuzzy number.

9.5.1.3 Profiling Kansei

To obtain the data necessary Kansei, two hundred eleven (211) respondents were asked, including researchers KE, Kanazawa experts and masters of traditional arts.

They were asked to evaluate the Kansei attributes of thirty (30) samples simultaneously to ensure the accuracy of personal judgments. These Kansei data were then used to generate the profiles Kansei the products tested.

9.5.1.4 Specification of Consumer Preferences

Each subject selected seven (7) attributes Kansei and seven (7) corresponding to attributes targets. Thus, it was possible to calculate the probability to find the product Om goals consistent with the Kansei attributes selected by the subjects.

9.5.1.5 Aggregate Performance of Prioritized Goals

The Kansei attributes were divided into three (3) priority levels. To add satisfaction levels in this structure were computed vector weights OWA for each priority level below the characteristic attitudinal Ωq.

Lower levels of satisfaction for the Kansei attributes induce the highest importance weights for attributes in lower priority levels lower.

9.5.2 Conclusion

For the authors, with the proposed methodology, the consumers could buy or select their favorite items according to your preference. The study focused only on Kansei attributes, and the resulting model suggests a consumer-oriented approach based on the Kansei profiles. The authors also conclude that the model is useful for the marketing area, particularly in the area of e-commerce, where recommendation systems have become a major area of research.

9.6 KE Type V: Virtual (Vike)

According to the author Nagamachi [10], virtual reality is an advanced technology capable of constructing a virtual space and provides an experience that is not possible to have in the real world.

The main characteristic of KE type V is the use of this technology, building a space and adapting it to the feeling of the user. Thus, it is possible to provide an experience to examine the approach to the design of a proposed Kansei expected by the user, as shown in Fig. 9.7.

It is very useful and efficient for products of large scale, because it is possible to analyze the virtual space before producing it physically. The steps of this kind of KE are based on the application made by Matsushita Electric Works and the Hiroshima University [11], as shown in Table 9.2.

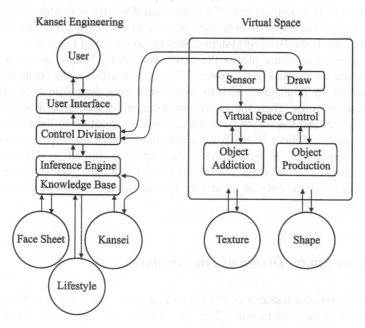

Fig. 9.7 System Structure of the VIKE. *Source* adapted from Nagamachi [10]

Table 9.2 Steps of KE type V according to Enomoto et al. [11]	Step 1	First, the users answer questions and give values for the Kansei words they want to be present in the design of the product/environment
	Step 2	The system used proposes an option based on the Kansei words of the users
	Step 3	If the graph shown on the screen is satisfactory, the data are processed by the virtual system
	Step 4	If the customers want any part to be modified, this modification is possible by changing the system module
	Step 5	The customers can change parts of the virtual product/environment
	Step 6	If the customers are satisfied, the virtual model of the product/environment is sent to the factory
	Step 7	All the parts of the product/environment will be sent to the customers' home in two (2) weeks and will be assembled in the presence of the customer

Initially, this type of KE was applied only to kitchens; later, it was expanded to other rooms in the house and became known in Japan as HousMall System, originally in the city of Osaka in 1999.

It is consisted of two (2) blocks of macro-systems computer: the main control is KE system, and the other is the virtual reality system.

The HousMall system consists of a main control system with user interface, the database image and Kansei words, the database of knowledge, the virtual reality system and finally the building system of design, as shown in Fig. 9.8.

The designed structure allows the consumer who wants to design a whole house put the family data in the system. The system asks for the future residents which part of the house they wish to start. So, it asks the residents to enter the system Kansei words. Then, the system proposes an option that can be viewed in a virtual space, and the future residents can see the solution to the environment from any angle. If there is any feature that does not please the residents, they can make changes in the system. After the improvements are made, the virtual space is approved by the users and the proposal is finalized. The future residents can still count on the help of an architect to define the details of the proposed environment.

9.6.1 Case Study: Design of Loudspeakers

The authors Artacho-Ramírez et al. [12] conducted a study on ways of representing products to be manufactured, how they influence the perception of the user/consumer and product success in the market.

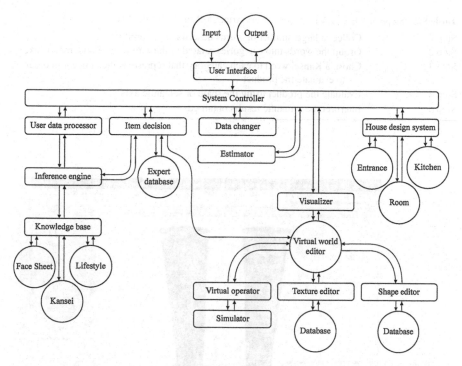

Fig. 9.8 The system structure of the HousMall. *Source* adapted from Nagamachi [10]

Thus, they conducted a study with different forms of representation of the product, among them the virtual space. For them, the difference of perception between the physical product and its representation decreases as it uses technologies that allow immersion and interactivity in real time and experience with the three-dimensionality.

In the case study of the authors Artacho-Ramírez et al. [12], the feelings of the user were observed concerning a loudspeaker and physical forms of representation (such as three-dimensional simulation). They use the semantic differential method and conducted the study in three (3) steps, which have features found in the types of Kansei until now operated, as shown in Table 9.3.

In this study, we used two (2) types of speakers: one traditional and another with different design features. The product was presented to the user via computer, and a reference element commonly known in size (CD) has been arranged next to the product to convey the concept of scale.

One type of analysis has been conducted giving the user the ability to interact with the product to modify the point of view of the object. To examine whether this possibility really have any interference in the user's feelings, the 3D models were converted into a type of format called Metastream, which can be played by almost all formats of Internet browsers (via a plug-in). This format allows a high quality and interactivity with 3D elements. Using the mouse, the user could move, rotate and vary the distance of the product and even test some of their functions (see Fig. 9.9).

Table 9.3 Steps of KE type VI for the authors Artacho-Ramírez et al. [12]

Step 1	Collect a large number of words describing the product
Step 2	Group the words into categories related to the same concept (semantic axis)
Step 3	Chose a Kansei word to each category that represents the concept in order to evaluate the product
Step 4	Evaluate the product using the scale or semantic axis
Step 5	Interpret the semantic results obtained

Fig. 9.9 Interface used to evaluate the product. *Source* adapted from Artacho-Ramírez et al. [12]

Another type of analysis, related to the three-dimensional representation, occurred with the use of a stereoscopic display which increases the degree of realism accentuating the immersion and the three-dimensional observer experience.

As a conclusion of the study, the authors discussed the contribution of the interaction between the user and the product before its release. This ensures that the product reaches your audience most likely to meet their needs and to arouse the desired feelings. The semantic differential was used to measure the influence of mode of representation in user perception.

When offered a product in a two-dimensional (photograph or image), there is a wide divergence of feelings and perceptions regarding the presence of the physical product. However, this divergence is minimized when using a 3D environment.

The possibility to observe the product of different points of view closer details move it or rotate it greatly improves their perception.

9.7 KE Type VI: Collaborative

According to Nagamachi [10], the KE type VI is constructed as a 'system design group' aided by an intelligent system and a database of Kansei words. This system is very useful to work together from various designers. In this type of KE is necessary to have a server with software that supports collaborative design, so designers can hear and see their partners working on the screen, which facilitates the exchange of ideas and information. The work is developed or adjusted using the database Kansei and following the instructions/suggestions of the system. After receiving the orders the company to which the product is developed, the designers check to see Kansei database and system.

In the system presented by Nagamachi [10], the designers begin the project with a meeting between members regarding the product's specifications and then depart for the development of the project. To do this, you must use the Kansei system and database. During the process, the designers talk to each other and exchange ideas and feelings. Figure 9.10 illustrates the structure of this collaboration system.

The collaborative design systems used the Internet to function. She has proved an efficient means of studies of user perception on products. The authors Kim et al. [13], through a study on the Internet, conclude that the evaluation method using virtual images of prototypes is an efficient way to analyze the impressions of the users of alternative products and presents a great cost benefit.

The authors Cho et al. [14] warn about how complicated it can be the analysis of user satisfaction, and that these high costs can be minimized with the implementation of the Internet as a means of interaction between designers and users. So, the KE type V can be very efficient in designing a new product, since the environment of the Internet and its technologies have been shown to be a possibility of obtaining the views of consumers in a short time and with minimum effort and resources.

9.7.1 Case Study: Interior of Car

In the study by the authors Kim et al. [13], it is highlighted the efficiency of using the evaluation method for Internet using virtual prototyping. This type of evaluation is a way to collect and analyze the views of several individuals on the alternatives of a product early in the process of its development. In this particular case, the authors Kim et al. [13] analyze the perception of the car interior. Figure 9.11 summarizes the steps of the study.

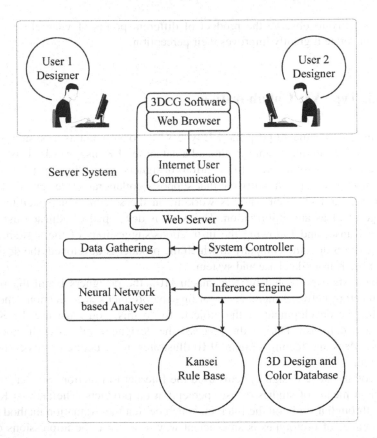

Fig. 9.10 Collaborative Kansei designing system. *Source* adapted from Nagamachi [10]

First, the design elements of interest were selected for the job, and then the alternatives were generated using virtual models. Two (2) experiments were conducted: in the virtual environment, using virtual models, and the Internet, through images of virtual models. The results of the two (2) experiments were analyzed and compared to test hypotheses generated by the search. Finally, a more efficient experimental technique has been discussed and suggested based on the results achieved.

9.8 Conclusion

As can be seen throughout the text, various tools and methods are used to apply a kind of KE. To get the image of a product expected by users, techniques, such as questionnaires, analysis of semantic differential, factor analysis, multidimensional scaling analysis, morphological analysis, fuzzy, neural network and linear regression, among others are used to understand the data that will be obtained. This feature is independent of the type of KE used, because within the same project can be seen

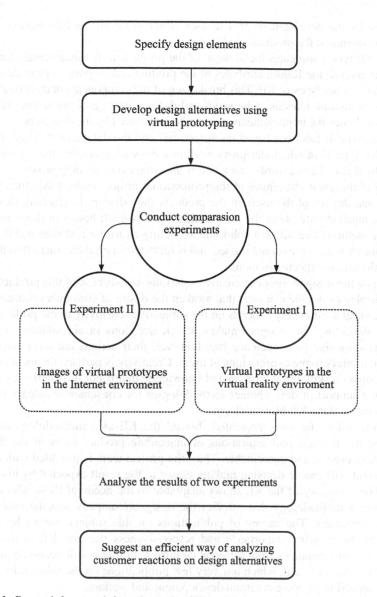

Fig. 9.11 Research framework from the study of Kim et al. [13]. *Source* adapted from Kim et al

the characteristics of different types of KE according to the needs of designers and users. Importantly, they are not mutually exclusive; more than one type of KE can be applied in the development of a product or support for the decision of a project.

You may notice that some procedures used are common to different types of KE. We can see a great interaction between them. The KE type I, for example, uses questionnaires and surveys to support the design decision. While for the initial procedure of KE type II, KE type I is the initial activity of the process followed by the development of databases which store all the information about the components of the products analyzed.

The KE type I stipulates the division of the product in its fundamental elements to better analyze the Kansei attributes of the product and its parts, a procedure that observed in other types of KE. This breakdown of the components in their best units specifies minimum relevance, the needs and desires of the user about this product, which facilitates the improvements that can be introduced by the designers.

In KE type II has been based on differential and the databases that are critical to the KE type V, in which designers work in a network accessing the information stored by them (Kansei words, stock colors and shapes) to develop project.

One of the great challenges of the professional project understands the expectations and desires of the users of the products they design. In teaching this profession, much debate about the need to achieve this goal; however, there are few tools or methods that allow a solid understanding as to the real user needs. This grace comes with professional trained and is reflected in products that often do not match the image expected of them.

Despite the designer propose creative solutions, the success of this product, that is, its display or not the elements that awaken the desire of consumers contemplate their needs and ultimately depends on the subjective sensitivity to the professional project develops. The designer makes design decisions in accordance with the understanding that your audience has, however, their actions are more informed on what it infers to the expectations of users. Creativity is perhaps the most prominent designer's skill, but it is an area of knowledge that is not fully understood and needs the support of development methodologies for consumer-oriented products from accurate data, KE Methodology.

In this sense, the work presented showed the KE as a methodology able to translate the feelings and aspirations of inaccurate product users in the design parameters precise and quantifiable. Thus, the project team is provided with information that will enable decision making closer to the result expected by his audience. The versatility of the KE shows adaptable to the needs of those who apply, so having a methodology that satisfies the designer/company and the users/customers/consumers. The survey of publications on this subject, shown by KEB, highlight the growing importance and responsiveness that the KE is found in industries and companies from different countries. In that is still necessary to promote this issue in Brazil, which has very few publications on the subject. Its diffusion is needed to promote common development and welfare.

References

1. Schütte S (2002) Designing feelings into products. integrating kansei engineering methodology in product development. Linköping Studies in Science and Technology. Linköping, Thesis No 946, ISBN:91-7373-347-4 ISSN:0280-7971
2. Nagamachi M (1995) Kansei engineering: a new ergonomic consumer-oriented technology for product development. Int J Ind Ergon 15:3–11

3. Jindo T, Hirasago K (1997) Application studies to car interior of kansei engineering. Int J Ind Ergon 19:105–114
4. Chuang MC, Ma YC (2001) Expressing the expected product images in product design of micro-electronic products. Int J Ind Ergon 27:233–245
5. Hayashi C (1976) Method of quantification. Toyokeizai, Tokyo
6. Llinares C, Page A (2007) Application of product differential semantics to quantify purchaser perceptions in housing assessment. Build Environ 4
7. Roy R, Goatman M, Khangura K (2009) User-centric design and kansei-engineering. CIRP J Manuf Sci Technol 1:172–178
8. Yan HB et al (2008) Kansei evaluation based on prioritized multi-attribute fuzzy target-oriented decision analysis. Inf Sci 178:4080–4093
9. El Marghani VGR et al (2011) Kansei engineering: metodologia orientada ao consumidor para suporte a decisão de projeto. Congresso Brasileiro de Gestão e Desenvolvimento de Produto
10. Nagamachi M (2002) Kansei engineering as a powerful consumer-oriented technology for product development. Appl Ergon 33:289–294
11. Enomoto N et al (1995) Kitchen planning system using kansei VR. Symbiosis of Human and Artifact, pp 185–190
12. Artacho-Ramírez MA, Diego-Mas JA, Alcaide-Marzal J (2008) Influence of the mode of graphical design representation on the perception of product aesthetic and emotional features: an exploratory study. Int J Ind Ergon 38:942–952
13. Kim C et al (2011) Evaluation of customer impressions using virtual prototypes in the internet environment. Int J Ind Ergon 41:118–127
14. Cho Y et al (2011) Development of a web-base survey system for evaluating affective satisfaction. Int J Ind Ergon 41:247–254

Chapter 10
Interaction Between Emotions and Mental Models in Engineering and Design Activities

Robert J. de Boer and Petra Badke-Schaub

Abstract The objective of this chapter is to describe the interaction between emotions and the preservation of ideas in engineering and design activities. Johnson-Laird's mental model construct of internal representations is utilized to define cognitive resistance. A description of the interaction between emotions and cognitive resistance is presented. A value function that forecasts a subject's behavior when confronted with stimuli that are contradictory to his mental model is derived from a reinforcement learning framework. An experimental study has been conducted to validate the theory. The relevant results, confirming the predictions of the value function, are presented.

10.1 Introduction

This chapter is about the emotional barriers that need to be overcome for engineers or designers to change their mind. As any engineer will know from his or her own experience, this does not always come naturally. I will argue that changing one's mind is regulated through emotions, building on Damasio's thoughts that emotions are essential to rational thinking and everyday behavior [1]. Consider for instance an engineer in aeronautics that has calculated the minimum height of a beam and is then told that this does not fit with the rest of the design. Or an industrial designer that has chosen a particular material for a new chair but finds this cannot be produced at the required cost. In both cases, we can expect that the person's reaction will probably be somewhat emotional, depending on how "passionate" he is about his original proposal. In some cases, like in the first example, the engineer will hopefully defend his position and uphold the beam dimensions (assuming he has done the calculations correctly). In the latter case, we may hope in the interest of economic feasibility that the designer is inspired to change his views.

R. J. de Boer (✉)
Amsterdam University of Applied Sciences, Weesperzijde 190, 1097 DZ, Amsterdam,
The Netherlands
e-mail: rj.de.boer@hva.nl

P. Badke-Schaub
Delft University of Technology, Landbergstraat 15, 2628 CE, Delft, The Netherlands

S. Fukuda (ed.), *Emotional Engineering vol. 2*, DOI: 10.1007/978-1-4471-4984-2_10,   149
© Springer-Verlag London 2013

10.2 Relation Between Emotions and Mental Models

Most engineers perform their activities under dynamic and uncertain circumstances. They are able to solve engineering problems and forward their work because they construct mental models of the world around them. Mental models reduce cognitive workload, and they are inherently stable in the face of contrary evidence, and so the assumptions underlying the mental model may diverge from reality. In the field of design, this discrepancy between the mental model and reality may continue for prolonged periods and could actually be justified in hindsight, for instance if the discrepancy is temporary. Similarly, the alignment of the individuals in group work is susceptible to the stability of the team member's mental models.

10.2.1 Mental Model Characteristics

The existence of mental models can be explained by the limited processing capacity of the human mind [2–4]. In real life, we are bombarded by a plethora of stimuli from which we need to distill some sense of coherence—yet minimize our cognitive load [5, 6]. Necessarily, mental models are simplifications of reality [7]. Mental models are parsimonious and holistic[1] which is in fact their utility, because as they contain less information, they are easier to work with and free resources for other cognitive processes [2, 5]. Mental models allow the integration of new perceptions with existing information to create an overall impression and retain a coherent view of reality [8, 5]. Mental models reduce complexity and therefore stress and allow us to "go beyond the information given" to give us a feeling of competence and control [9, p. 401].

Based on work by Johnson-Laird and other cognitive psychologists as cited above, we can conclude that mental models are simplified, holistic, resilient internal representations of reality in working memory that are created and demised subconsciously, that guide our thinking and action and that free up cognitive resources. The construct seems to share some or all of its defining characteristics with similar constructs like "schemata," "schema," "scripts," "frames" and "situation awareness." However, the mental model construct utilized in this chapter differs from the construct by the same name that is commonly applied in the human factors community on four points [10–12]: it is available in working memory, it is created and terminated subconsciously, it is a holistic representation, and it is relatively resilient.

---

[1] Meaning that the model only makes sense as a whole; the constituent parts are unimportant and not easily accessible.

10.2.2 Discrepancies Between Reality and a Mental Model

On the down side, mental models also entail a risk of conjuring too much in our imagination [8]. Mental models are inherently stable in the face of contrary evidence because they guide our attention, our thinking and thus our actions [8, 13–16]. They are essentially built to pursue a given goal based on data extracted from the environment. As a result, essential features of a problem are overemphasized, whereas the peripheral data may be neglected [17]. We risk missing new information [15] and are inhibited to try alternative approaches even when warranted [18]. The simplification of reality is not a conscious act but an autonomous part of the human information processing function [19]. Crucial to the theory of mental models is that we usually represent only one possibility in our mind when reasoning, even when multiple options are available [14].

The creation and termination of a mental model is an autonomous, involuntary process that is not subject to self-reflection:

> The process of construction is unconscious, but it yields a representation, and this mental model enables us to draw a conclusion, by another unconscious process. [...] In general, the world in our conscious minds is a sequence of representations that result from a set of processes, and the world in our unconscious mind is the set of processes themselves [14, p. 53]

Individuals are not aware of the creation or termination of mental models and cannot control these processes, just as we are not aware of many high-level processes in problem solving and creative thinking [20]. The mental model itself is accessible for cognitive thought because it is in working memory, and it is used as a building block in subsequent steps [21].

The discrepancy between reality and a mental model has been described by many authors (particularly from the human factors community) in different terms (e.g., functional fixity, cognitive mismatch, failure to apprehend, selective attention deficit, fixation, perceptual set, plan continuation bias, illusion of attention, tunneling, perseveration syndrome, lack of cognitive flexibility, cognitive lockup). Note that many of these terms imply erroneous behavior on the part of the individual, whereas the deviation of the mental model from reality is actually an inherent consequence of its functionality. For the purpose of the current study, a new term for this construct is proposed that better matches the phenomena under consideration: *cognitive resistance*. In this work, cognitive resistance is defined as the capacity to endure contradictory stimuli from the environment until reflection on the assumptions underlying the mental model. During cognitive resistance, new perceptions are ignored or interpreted in such a way that they fit the existing mental model.

10.2.3 The Role of Emotions

Authors like Donald Schön [22] suggest that emotions may contribute to the resolution of cognitive resistance. In **Reflective Practice**, the designer reflects on his own work to enable progress after "pleasing [...] or unwanted" surprises. He suggests that surprise is the trigger for engaging in a reflective mode of thinking (i.e., demise of the existing mental model):

> When intuitive, spontaneous performance yields nothing more than the results expected for it, then we tend not to think about it. But when intuitive performance leads to surprises, pleasing and promising or unwanted, we may respond by reflection-in-action [22, p. 56].

From this description follows that according to Schön, surprise is a necessary condition for reflection, and therefore, "being surprised" reduces cognitive resistance. Other emotional responses such as anger similarly seem to have an effect on cognitive resistance:

> Participants in the game not infrequently became attached to a particular reading of the prototype, and treated an alternative reading as a threat, which provoked an angry and defensive reaction [23, p.145].

Other authors, building on Schön's work, have corroborated the effect of different emotions on mental model preservation in design. Akin [24] describes the surprise in what he calls the "Aha! Response" when existing ideas are successfully challenged. Kleinsmann [25, pp. 170–179] presents an example where reflection is inhibited, and current thinking persists due to anger. Each of these cases highlights that conflicts may arise in team settings due to the challenges to the mental models of the individual team members. In some cases, these emotions may initially seem disruptive, but actually turn out to be fruitful (cf. [26]), because they support the development of the team mental model [5, 27].

Mandler's Theory of Discrepancy and Interruption [28] replicates Schön's thinking, reserving a central role for interruptions. Mandler proposes that "interruption is a sufficient and necessary condition for the occurrence of autonomic arousal." He writes:

> "A new input that activates a new schema may be interrupting, if the new schema is incompatible with the old, if it contradicts the operation of the old structure or, more generally, if it provides evidence that it [...] cannot be assimilated by the existing structures".

In contrast to Reflective Practice, Mandler suggests that the interruption may lead to different, specific types of emotion, depending on "factors other than the interruption itself." The emotions can be either positively or negatively valenced. Mandler does not qualify the effect of different interruptions according to elicited emotion types.

The **Somatic Marker Hypothesis** has been influential in highlighting the role of affect in rational decision making. The hypothesis was posited to account for the role of emotion in the decision-making process [29]. This theory states that previous emotional experiences are reactivated whenever individuals face a situation that has previously been "categorized" (i.e., marked somatically). This then supports future decision making in a similar context [30]. Research findings indicate that an emotional response is triggered before conscious reflection.

The **Communicative Theory of Emotions** complements Johnson-Laird's mental model theory [14] cited earlier. According to the communicative theory of emotions, subconscious evaluations trigger emotions as signals to direct attention, mobilize the body, and to prepare for appropriate behavior [31, 32]. Emotions set the brain into specific states to coordinate our multiple goals, given our limited intellectual resources, and initiate reasoning [14, p. 87]. The communicative theory of emotions proposes that emotions are triggered subconsciously to activate consciousness, implying a role in reflection and demise of the mental model. The type of emotion influences our intentions.

The **Dual Process Theory of Reasoning** [33] has attracted much research within psychology and according to Johnson-Laird [21] aligns with the mental model theory of reasoning. In dual process terms, system 1 and system 2 are in conflict if the inherent resistance of the mental model leads to a sufficiently large mismatch with reality. Different authors have shown that affective processing is engaged if system 1 and system 2 are in conflict. This is even the case when deliberate thought processes are restrained, and individuals respond intuitively, that is, they reject the challenge to the mental model [34, 35]. There is no activation of these specific brain regions without conflict, that is, if the conflicting stimuli are not perceived because they do not surpass the perception threshold. Baumeister and Masicampo [36] have recently proposed that conscious thought influences behavior only indirectly. They suggest that emotions "serve to stimulate conscious reflection on past and future events" and serve as feedback to learn from past actions, as a bridge between subconscious (system 1) and conscious thought (system 2). For instance, negative emotions seem to stimulate counterfactual and detail-oriented thinking, and regret and guilt promote learning [37].

The **functional nature of emotions** reflects the paradigm of evolutionary psychologists (e.g., [38]). They suggest that emotions are a flexible response to stimuli from the environment. A large discrepancy between reality and an existent mental model can be life threatening and may be signaled by emotions. Emotions are considered by evolutionary psychologists to be functional in terms of our evolutionary, biological and/or social survival: they stimulate behavior that allows us to pass on our genes, to stay alive at least long enough to reproduce and care for our offspring and to maintain social cohesion in support of our fitness and survival [39]. Emotions enable a more flexible response to particular stimuli than reflexes or habits permit because they are goal oriented rather than action specific [40]. The elicitation of emotions costs energy and so from a purely biological point of view would appear to require clear benefits.

Most authors, starting with Charles Darwin [41], at least agree that emotions were *originally* functional for the survival of the species. Frijda [42] similarly agrees that "emotions are largely viewed as adaptively useful,[2] or at least have been in the evolutionary past." Some emotions may now be "mere obsolete

---

[2] In the meaning of "functional for survival".

remnants," although "by and large, emotions and emotional actions are still generally adaptive in about the original sense." Frijda goes on to state that other emotions like joy, excitement, curiosity and grief are functional in the non-adaptive sense (i.e., functional but not necessary for the individual's survival). Oatley [43] extends the label of functionality to most (if not all) emotions, stating that emotions are predominantly aimed at mediating social relationships and are deemed functional in the context of our interdependency with other human beings.

In contrast, evolutionary psychologists (e.g., [38]) consider even these emotions "adaptive," because they promote the reproduction of genes in the self, children and relatives (instead of just the survival of individuals). Those able to "read the intentions of others and engage their solicitude would be more likely to survive and prosper" [44]. Emotions support "social survival" by helping to form and maintain social relationships and a social position relative to others [45]. Even though all or at least most emotion types can be deemed functional from a biological perspective, this of course does not imply that in all specific instances emotions are useful or advantageous. This is dependent upon the specific circumstances and the emotion regulation that is embedded in this specific occurrence of the emotion [46].

The **model of Intentional Dynamics** [47] explains how both error and surprise are inevitable as individuals struggle to keep pace with changes in a dynamic environment. The authors consider this model a first step to merge the theories of cognitive psychology with the practical concerns of human factors. The model shows how interaction with the environment (ecology) is driven by assumptions about reality (belief). The authors stress that beliefs are generated by the integration of continuous interaction with the environment over time rather than discrete memories of specific moments. The interface "compares or mediates" the consequences of an action with intentions and expectations. The interface is able to trigger a modification to the beliefs through a "surprise" or to discount the information. The authors incorporate in the "interface" both the *media* of perception (e.g., stimuli and field of view) and the *mechanisms* of perception (e.g., receptors, short-term sensory store and perceptual encoding).

In summary, from the literature, ample evidence is available to suggest that emotions play a moderating role in the demise of a mental model. Both the elicitation of emotions and the type of emotion (if elicited) seem to be relevant.

10.3 Value Function

In the previous section, it was suggested that emotions play a moderating role in the demise of a mental model. This can be modeled using a reinforcement learning framework, as this is a broadly applicable tool to frame the problem of learning from interaction to achieve a goal [48]. The use of the reinforcement learning framework in the context of cognitive resistance is justified by the proposed adaptive nature of stimulus matching and the ability of the framework to incorporate the interaction of emotion types. The framework has been applied to model similar types of psychological processes in the past [49,50].

10.3.1 Reinforcement Learning Framework

A reinforcement learning framework assumes a learner/decision maker (called *agent*) that interacts with an *environment* (comprising everything outside the agent). The agent and environment interact continually, the agent selecting actions and the environment responding to those actions by presenting new situations (or system states) and giving rise to rewards. In reinforcement learning, it is presumed that the actions of the agent are driven by its expectations of future rewards and the aim to maximize these. The real future returns are not initially known by the agent until it gains experience about the probabilities and rewards associated with state transitions through repetitive trials and updates its expectations accordingly. The convergence between the real rewards and the agent's estimation of the rewards constitutes "learning" in the reinforcement learning framework [48].

In the current context, the agent is proposed to be the subconscious controller of the process step that decides whether to maintain (preserve) or demise the mental model, so-called *stimulus matching*. This controller maintains an optimum balance (the goal) between ignoring irrelevant stimuli (thereby saving resources) and acting (i.e., reflecting) upon significant stimuli in the interest of survival. The agent is expected to adapt its behavior over time, depending on the perseverance of contradictory stimuli. The framework requires that the reward mechanism is external to the agent [48, p. 53]. The subconscious controller that matches stimuli should be independent of the appraisal mechanism that triggers emotional responses. This is justified for the current context by the suggestion that these processes occur in separate brain areas: the limbic system for appraisal and emotional response, and the cortex for stimuli matching [51]. The action choices available in stimulus matching (the agent) are (1) to reflect on the stimulus that contradicts the assumption underlying the current mental model or (2) to ignore the stimulus and preserve the mental model. The adaption of agent's behavior depends on the feedback (state and reward signals) it receives from the environment. The state signal is the stimulus that represents the discrepancy between reality and the mental model. The emotion type of the elicited emotional response represents the reward mechanism; negatively valenced emotions being valued as a negative reward (i.e., penalty). The reinforcement learning framework in the context of cognitive resistance is shown in Fig. 10.1.

The agent is expected to behave so as to maximize its expected future rewards. A distinction needs to be made between the *real* returns that actually follow from the agent's actions and the agent's *estimation* of the expected returns. Only the latter drives the agent's behavior, but the former drives the rewards or penalties the agent experiences and therefore indirectly allows the agent to update the estimation of the expected returns.

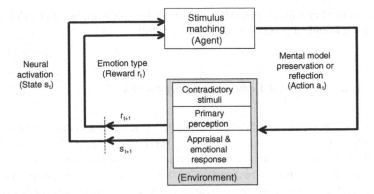

Fig. 10.1 Reinforcement learning framework for cognitive resistance, showing emotion as a penalty or reward for preserving a mental model

10.3.2 Real Returns

The real return for the agent depends on the actions he selects: either to reflect on the stimulus or to preserve the mental model at each time step[3] that he encounters a contradictory stimulus. It is proposed that reflection results in a significant negative reward (i.e., a penalty) due to the additional burden on the mental processes. Additionally, it is proposed that negatively valenced emotions constitute a negative reward and positively valenced emotions represent a positive reward. If the agent chooses to reflect, the episode is terminated. If the agent chooses to preserve the mental model, the episode continues, as cognitive resistance is defined by the capacity to endure contradictory stimuli until reflection.[4] Therefore, the real future return (R_t) for the whole episode from a time step (t) until reflection at some time step (k) in the future and preservation of the mental model at all time steps until then is defined as follows[5]:

$$R_t = r^{\text{reflect}} + \sum\nolimits_{j=0}^{k} r^{\text{ignore}}(j) \tag{10.1}$$

Under the conditions that discounting can be disregarded[6] and

$$k, j \in \mathbb{N}; \ k \geq j \geq 0 \tag{10.2}$$

---

[3] An analogous analysis is warranted in the context of continuous time.

[4] In real life, contradictory stimuli may—and often do—extinguish, of course.

[5] An important assumption is that all of the relevant history of the environment is represented in each state signal, so that these have Markov properties, and the reinforcement learning framework represents a Markov decision process.

[6] This simplification allows us to disregard the non-contradictory stimuli that are usually interspaced between the contradictory stimuli and is allowable for the qualitative discussion in this section.

$$a_j \in \{\text{ignore, reflect}\} \tag{10.3}$$

$$\forall (j < k) : a_j = \text{ignore} \tag{10.4}$$

$$a_k = \text{reflect} \tag{10.5}$$

where

- $R_t\left(k, r^{\text{ignore}}(j)\right)$ is the real future return at time t for the whole episode.
- $r^{\text{reflect}} \ll 0$.
- k is the number of time steps with contradictory state signals until reflection.
- j is a counter for time steps with contradictory signals.
- $r^{\text{ignore}}(j) < 0$ for negative emotions and $r^{\text{ignore}}(j) > 0$ for positive emotions.
- a_j is the action taken at time step j, the action is to ignore the contradictory stimulus every time step except the last (k).

Equation 10.1 shows that the real future return (R_t) at time step (t) is dependent upon the future actions (or policy) of the agent in that the agent may choose to reflect at any time step with a contradictory stimulus (k) in future.

10.3.3 Returns Expected by the Agent

In reinforcement learning, the agent is not initially aware of the *real* future rewards. The agent is assumed to update its *expectations* of the future returns at each time step through the reward or penalty that is experienced, without necessarily having a model of the environment's dynamics. This constitutes a so-called temporal-difference learning situation [48]. In temporal-difference learning, the current estimate of future rewards is updated by the reward or penalty that is incurred by the action taken at each time step. Mathematically, the update of the estimate of future rewards (or *value function*[7]) is defined as follows:

$$V(s_t) \leftarrow V(s_t) + \alpha \left[r_{t+1} + \gamma V(s_{t+1}) - V(s_t) \right] \tag{10.6}$$

where

- $V(s_t)$ is the current estimate of the value function in state s_t.
- α is a constant indicating the learning rate at which the current estimated value function is updated $(0 < \alpha < 1)$.
- r_{t+1} is the immediate reward at time $t + 1$ following the action a_t in state s_t.
- γ is the discount factor for future rewards in comparison with immediate rewards $(0 < \gamma \leq 1)$.

---

[7] Following [48], we use the term "value function" to denote the expected future rewards if the agent follows a particular set of consecutive actions (a so-called policy).

From Eq. 10.1, it follows that in the context of cognitive resistance, the expected return is independent of the state (s_t), if the time steps with contradictory signals only are regarded. By substituting $V(j)$ for $V(s_t)$ and $V(j + 1)$ for $V(s_{t+1})$ and ignoring discount $(\gamma \cong 1)$, it follows that as long as the agent does not reflect (and therefore, the episode does not terminate, i.e. $j < 1$)

$$V_{\text{ignore}}(j + 1) = V_{\text{ignore}}(j) + \frac{\alpha}{1 - \alpha} r^{\text{ignore}}(j + 1) \tag{10.7}$$

$$\forall j : s_j = \text{contradictory stimulus} \tag{10.8}$$

The agent's expected total return for each of his choice of actions $(a_j \in \{\text{ignore, reflect}\})$ is therefore

$$\forall j > 0 : V_{\text{ignore}}(j) = V_{\text{ignore}}(j - 1) + \frac{\alpha}{1 - \alpha} r^{\text{ignore}}(j) \tag{10.9}$$

$$V_{\text{reflect}}(j) = r^{\text{reflect}} \tag{10.10}$$

where

- $V_a(j)$ is the current estimate of the value function for action a at time j; $V_{\text{ignore}}(0)$ is as yet undefined.
- j is a counter for time steps with contradictory signals only; $j \in \mathbb{N}$.
- α is a constant indicating the learning rate at which the current estimated value function is updated $(0 < \alpha < 1)$.
- $r^{\text{ignore}}(j)$ is the immediate reward at time j following the action to ignore at time $j-1$, varying with the time steps (j).
- r^{reflect} is a significant negative constant reflecting the penalty of reflection $(r^{\text{reflect}} \ll 0)$.

The agent is expected to choose a sequence of actions (a policy) that is *greedy* and therefore predominantly aims to maximize total return [48]:

$$V_{\text{ignore}}(j) > V_{\text{reflect}}(j) \rightarrow a_j = \text{ignore} \tag{10.11}$$

$$V_{\text{ignore}}(j) < V_{\text{reflect}}(j) \rightarrow a_j = \text{reflect} \tag{10.12}$$

The agent will choose to ignore the contradictory stimulus at $j = 0$ (necessary for cognitive resistance to occur) under the condition that

$$V_{\text{ignore}}(0) > V_{\text{reflect}}(0) = r^{\text{reflect}} \rightarrow a_0 = \text{ignore} \tag{10.13}$$

The agent knows by previous experience that reflection will result in a significant negative reward due to the additional burden on its mental processes $(r^{\text{reflect}} \ll 0)$. For Eqs. 10.11 and 10.13 to be true, the agent may assume — based on previous experience — that the contradictory stimulus is temporary and will go away. As the contradictory stimuli persevere, the agent will update its expectancy for future rewards with the repetitive rewards and penalties that it is experiencing according to Eq. 10.9. As indicated above, negatively valenced emotions constitute a negative reward: $r^{\text{ignore}}(j) < 0$; positively valenced emotions represent a positive reward $r^{\text{ignore}}(j) > 0$. At some point, the expected penalties for preservation may exceed the penalty for reflection (i.e., Eq. 10.12 is true), and the agent will choose reflection instead of preservation. The value of $r^{\text{ignore}}(j)$ in Eq. 10.9 is proposed to depend on the emotion type that is elicited during cognitive resistance. That is, the speed in which the agent's expected return is adapted is sensitive to the emotional penalty (or reward) which he experiences each time he chooses to ignore the contradictory stimulus and maintain his mental model.

10.3.4 Emotion Types

The speed of learning in the reinforcement learning framework has been shown to be dependent upon the emotional penalty (or reward) which the agent experiences each time he chooses to ignore the contradictory stimulus and maintain his mental model. The penalties are discussed for the most relevant emotions in this section.

Joy is defined as the positively valenced emotional response to an actual event. Synonyms include pleasure, contentment, cheerfulness, delight, elation, euphoria, gladness and happiness [52]. Joy[8] is elicited after an event is appraised as pleasurable. Positive emotions are generally associated with the ability to switch attention [53] and to *broaden-and-build*[9] [54]. All positive emotions include "acceptance" of a situation [55, p. 85] — emphasized for instance in satisfaction and relief. "Positive affect [...] tends to facilitate a receptive and holistic, rather than [an] active, analytical mode of attention" [55, p. 73].

Surprise is defined as the non-valenced emotional response to an unexpected actual event. It includes attentional activity, novelty and unexpectedness. Surprise is elicited after a violation of expectancy [56]. It is associated with a tendency to attend to the cause of the surprise, bringing the event into consciousness, as described in much of the design literature.

Distress is defined as negative affect about an undesirable event. Synonyms include depressed, distressed, displeased, dissatisfied, distraught, feeling bad, feeling uncomfortable, grief, regret, sad, unhappy, etc [52]. Distress (or sadness) is not directed at an agent, but is a response to an event [52].

Anger is a negatively valenced compound response to a consequence of an actual event and the attribution of its cause to the action of an other agent. Synonyms include annoyance, exasperation, fury, indignation, irritation, offended, etc [52]. Anger's adaptive function is to oppose, overcome and master obstacles [57]. Anger is aimed at repelling the other agent [57].

---

[8] Joy will be used as the general term to describe the positively valenced emotions in this chapter.

[9] That is, "to broaden thought-action repertoires and lead to actions that build enduring personal resources" [54].

Remorse is the negatively valenced compound response to a consequence of an actual event and the attribution of its cause to the action of self. It is a mixed emotion of shame and distress. Synonyms include penitent, self-anger, embarrassment etc [52]. Remorse is focused on blaming oneself, trying to undo the situation, and even wishing to disappear [58].

10.3.5 Prediction Using the Value Function

In this section, a dynamic model for cognitive resistance was developed using a reinforcement learning framework to allow the analysis of the episodic nature of mental model preservation, in which emotion type has an effect on cognitive resistance and a cumulative effect of contradictory stimuli on the probability of reflection is assumed. From the descriptions of the emotions given above, it follows that the rewards associated with each emotion type is as follows:

- Joy is positively valenced, therefore associated with a (positive) reward;
- Distress is generally moderately negatively valenced and therefore associated with a moderate penalty;
- Anger is generally strongly negatively valenced and associated with a high penalty;
- Remorse is also strongly negatively valenced and associated with a high penalty;
- Surprise is not associated with a penalty or reward but with a tendency to attend to the cause, bringing the event into consciousness; and
- If no emotion is elicited, then the associated reward or penalty is close to zero, and there is limited tendency to attend to the cause.

Based on these estimations and the value function defined by Eqs. 10.9 and 10.10, it follows that emotion-type biases stimulus matching, such that surprise leads to immediate reflection, joy inhibits reflection, distress leads to a slow rate of reflection, anger leads to rapid reflection, remorse leads to a moderate rate of reflection, and reflection is inhibited if no emotion is elicited.

10.4 Experimental Validation

In this chapter, we have proposed that emotions are instrumental in signaling a discrepancy between reality and an existing mental model. We have predicted the effect of different emotion types using a reinforcement learning framework. These predictions have been tested empirically (see [59] for a full account), utilizing a number reduction task. This task was used by Wagner and colleagues [60] to identify reflection through an abrupt change in behavior and was originally developed by Thurstone and Thurstone. In total, 81 engineers from universities and industry participated in the experiment.

The results from the study show that emotions are elicited in the course of cognitive resistance. The predictions for the interaction of components of cognitive resistance have been validated:

- Joy was not observable directly before reflection, as predicted by the value function;
- Surprise largely led to immediate reflection;
- Distress was shown to loop back to itself, therefore leading to a low rate of reflection;
- Anger leads to reflection in one-third of the cases;
- Remorse leads to a moderate rate of reflection; and
- Reflection is inhibited if no emotion is elicited despite the contradictory stimuli.

In Fig. 10.2, the emotions are shown that are elicited during the course of the experiment for consecutive contradictory stimuli. These results lend credibility to the suggestion that a reinforcement learning framework as introduced in this chapter can be utilized to model cognitive resilience.

10.5 Conclusion

The prime objective of this chapter was to describe the interaction between emotions and the preservation of ideas in engineering and design activities. Three main points have been presented. Firstly, the interaction between emotions and internal representations (called mental models) has been discussed. That is, a stimulus that contradicts an existing mental model will trigger the elicitation of an emotion. The type of elicited emotion is dependent on the circumstances but also the

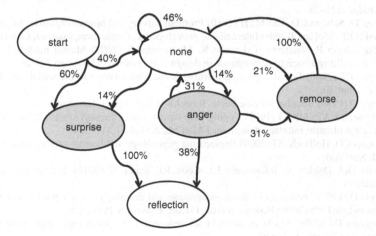

Fig. 10.2 Exit probabilities for emotions at consecutive contradictory stimuli (frequencies <2.5 % have been ignored; sum of exit probabilities = 100 %)

personal disposition of the subject at that time. Secondly, the valence of the emotion (i.e., whether it is pleasurable or not) can be considered a cost or a benefit for the subject. Therefore, in this chapter, a value function for the interdependency of emotions and mental models has been derived. This value function forecasts a subject's behavior when confronted with stimuli that are contradictory to his mental model. Finally, results of an experiment which corroborates the predictions of the value function for the interaction between emotions and a mental model have been presented.

The findings of the current work contribute to the science of design methodology by providing a theoretical and empirical foundation for the study of the resistance of mental models and emotions. The interaction between emotions and reflection that have been suggested by Schön has been validated. This study also contributes to previous attempts to relate emotions to design performance, by suggesting that the relationship may not be a direct one. Rather, the findings of this study suggest that emotions are influential in achieving reflection, and reflection may or may not improve performance. Practical examples of cognitive resistance in design can help designers and managers to recognize the subconscious and cognitive processes that are involved in designing complex systems.

References

1. Damasio AR (1994) Descarte's error: emotion, reason and the human brain. New York, Grosset/Putnam
2. Johnson-Laird PN (2006b) Mental models, sentential reasoning, and illusory inferences. Adv Psychol, pp 27–51
3. Miller GA (1956) The magical number seven, plus or minus two: some limits on our capacity for processing information. Psychol Rev 63:81–97
4. Newell A, Simon HA (1972) Human problem solving. Prentice-Hall Englewood Cliffs, NJ
5. Boos M (2007) Optimal sharedness of mental models for effective group performance. CoDesign 3:21–28
6. Nokes TJ, Schunn CD, Chi MTH (2010) Problem solving and human expertise. In: Peterson P, Baker EL, McGaw B (eds) International encyclopedia of education. Elsevier, Oxford
7. Badke-Schaub P, Neumann A, Lauche K, Mohammed S (2007) Mental models in design teams: a valid approach to performance in design collaboration? CoDesign 3:5–20
8. Higgins ET (2000) Social cognition: learning about what matters in the social world. Eur J Soc Psychol 30:3–39
9. Dörner D (1999) Bauplan für eine Seele. Rowohlt Verlag GmbH, Hamburg
10. Rasmussen J, Vicente KJ (1989) Coping with human errors through system design: implications for ecological interface design. Int J Man Mach Stud 31:517–534
11. Wickens CD, Hollands JG (2000) Engineering psychology and human performance. Prentice Hall, New York
12. Woods DD, Dekker S, Johannesen LJ, Cook RI, Sarter N (2010) Behind human error. Ashgate
13. Dörner D (1997) The logic of failure: recognizing and avoiding error in complex situations
14. Johnson-Laird PN (2006) How we reason. Oxford University Press, USA
15. Schraagen JM (2009) Macht en onmacht der gewoonte (The power and weakness of habits). Universiteit Twente, Enschede
16. Stempfle J, Badke-Schaub P (2002) Thinking in design teams: an analysis of team communication. Des Stud 23:473–496

17. Besnard D, Greathead D, Baxter G (2004) When mental models go wrong: co-occurrences in dynamic, critical systems. Int J Hum Comput Stud 60:117–128
18. Cardoso C, Badke-Schaub P, Luz A (2009) Design fixation on non-verbal stimuli: the influence of simple vs rich pictorial information on design problem solving. In: Proceedings of the ASME 2009 international design engineering technical conferences and computers and information in engineering conference. San Diego, California
19. Dijksterhuis A, Nordgren LF (2006) A theory of unconscious thought. Perspectives Psychol Sci 1:95
20. Bargh JA, Chartrand TL (1999) The unbearable automaticity of being. Am Psychol 54:462–479
21. Johnson-Laird PN (2010) Mental models and human reasoning. Proc Natl Acad Sci 107:18243–18250
22. Schön DA (1983) The reflective practitioner. Basic Books, New York
23. Schön DA (1992) Designing as reflective conversation with the materials of a design situation. Knowl Based Syst 5:3–14
24. Akin C (2008) Frames of reference in architectural design: analyzing the hyper-acclamation (A-h-a-!). Carnegie Mellon University
25. Kleinsmann M (2006) Understanding collaborative design. School of Industrial Design Engineering. Delft University of Technology
26. Tuckman BW, Jensen MAC (1977) Stages of small-group development revisited. Group Organ Manage 2:419–427
27. Badke-Schaub P, Goldschmidt G, Meijer M (2010) How does cognitive conflict in design teams support the development of creative ideas? Creativity Innovation Manage 19:119–133
28. Mandler G (1984) Mind and body: Psychology of emotion and stress. WW Norton, New York
29. Bechara A, Damasio AR (2005) The somatic marker hypothesis: a neural theory of economic decision. Games Econ Behav 52:336–372
30. Bechara A, Damasio H, Damasio AR (2000) Emotion, decision making and the orbitofrontal cortex. Cereb Cortex 10:295
31. Johnson-Laird PN, Oatley K (2008) Emotions, music and literature. In: Lewis M, Haviland-Jones JM, Feldman Barrett L (eds) Handbook of emotions, 3rd edn. The Guilford Press, New York
32. Oatley K, Johnson-Laird PN (1996) The communicative theory of emotions: empirical tests, mental models, and implications for social interaction. In: Martin LL, Tesser A (eds) Striving and feeling: interactions among goals, affect, and self-regulation. Routledge
33. Stanovich KE, West RF (2000) Individual differences in reasoning: implications for the rationality debate? Behav Brain Sci 23:645–665
34. de Neys W, Franssens S (2009) Belief inhibition during thinking: not always winning but at least taking part. Cognition 113:45–61
35. Pessoa L, Japee S, Sturman D, Ungerleider LG (2006) Target visibility and visual awareness modulate amygdala responses to fearful faces. Cereb Cortex 16:366–375
36. Baumeister RF, Masicampo EJ (2010) Conscious thought is for facilitating social and cultural interactions: How mental simulations serve the Animal-Culture interface. Psychol Rev 117:945–971
37. Baumeister RF, Vohs KD, Nathan Dewall C (2007) How emotion shapes behavior: feedback, anticipation, and reflection, rather than direct causation. Pers Soc Psychol Rev 11:167
38. Tooby J, Cosmides L (2008) The evolutionary psychology of the emotions and their relationship to internal regulatory variables. In: Lewis M, Haviland-Jones JM, Feldman Barrett L (eds) Handbook of emotions, 3rd edn. The Guilford Press, New York
39. Turner JH (2000) On the origins of human emotions: a sociological inquiry into the evolution of human affect. Stanford University Press, Stanford
40. Rolls ET (2007) Emotion elicited by primary reinforcers and following stimulus-reinforcement association learning. In: Coan JA, Allen JJB (eds) Handbook of emotion elicitation and assessment. Oxford University Press, USA

41. Darwin C (1872) The expression of emotion in man and animals
42. Frijda NH (2008) The Psychologist's point of view. In: Lewis M, Haviland-Jones JM, Feldman Barrett L (eds) Handbook of emotions, 3rd edn. The Guilford Press, New York
43. Oatley K (2007) On the definition and function of emotions. Soc Sci Inf, 46
44. Hrdy SB (2007) Evolutionary context of human development: the cooperative breeding model. In: Salmon C, Shackelford TK (eds) Family relationships: an evolutionary perspective. Oxford University Press, USA
45. Fischer AH, Manstead ASR (2008) Social functions of emotion. In: Lewis M, Haviland-Jones JM, Feldman Barrett L (eds) Handbook of emotions, 3rd edn. The Guilford Press, New York
46. Gross JJ (2008) Emotion regulation. In: Lewis M, Haviland-Jones JM, Feldman Barrett L (eds) Handbook of Emotions, 3rd edn. The Guilford Press, New York
47. Flach JM, Dekker S, Stappers PJ (2008) Playing twenty questions with nature: reflections on quantum mechanics and cognitive systems. Theor Issues Ergon Sci 9:125–154
48. Sutton RS, Barto AG (1998) Reinforcement learning, MIT Press, Cambridge
49. Holroyd CB, Coles MG (2002) The neural basis of human error processing: reinforcement learning, dopamine, and the errorrelated negativity. Psychological review, 109(4):679
50. O'Doherty JP, Dayan P, Friston K, Critchley H, Dolan RJ (2003) Temporal difference models and reward-related learning in the human brain. Neuron, 38(2):329
51. Wickens A (2009) Introduction to Biopsychology. Prentice Hall, Englewood Cliffs
52. Ortony A, Clore GL, Collins A (1988) The cognitive structure of emotions. Cambridge University Press, Cambridge
53. ISEN, A. M. (2008) Some Ways in Which Positive Affect Influences Decision Making and Problem Solving. IN LEWIS, M., HAVILAND-JONES, J. M. & FELDMAN BARRETT, L. (Eds.) Handbook of Emotions. Third ed. New York, The Guilford Press
54. Fredrickson BL, Cohn MA (2008) Positive emotions. In: Lewis M, Haviland-Jones JM, Feldman Barrett L (eds) Handbook of emotions, 3rd edn. The Guilford Press, New York
55. Frijda NH (2007) The laws of emotion. Lawrence Erlbaum Associates, Inc., Mahwah, NJ
56. Lewis MD (2008b) The emergence of human emotions. In: Lewis M, Haviland-Jones JM, Feldman Barrett L (eds) Handbook of emotions, 3rd edn. The Guilford Press, New York
57. Lemerise EA, Dodge KA (2008) The development of anger and hostile interactions. In: Lewis M, Haviland-Jones JM, Barrett LF (eds) Handbook of emotions, 3rd edn. The Guilford Press, New York
58. Lewis MD (2008a) Self-conscious emotions: embarrassment, prode, shame and guilt. In: Lewis M, Haviland-Jones JM, Feldman Barrett L (Eds.) Handbook of emotions, 3rd edn. The Guilford Press, New York
59. de Boer RJ (2012) Seneca's error: an affective model of cognitive resilience *Industrial Design Engineering*. Delft University of Technology, Delft
60. Wagner U, Gais S, Haider H, Verleger R, Born J (2004) Sleep inspires insight. Nature 427:352–355

Chapter 11
Emotional Quality Inspection for Revealing Product Quality

Teruaki Ito

Abstract Product quality is one of the critical issues to be competitive in the global marketplace. In addition, food production companies are strongly required to control high quality of products in their production lines. Contaminations such as metallic, plastic or organic substances are often harmful to human body and should be eliminated completely. Therefore, machine inspections, such as X-ray or fluorescence spectrum methods, are effectively used to eliminate these contaminating substances. However, the final inspections are taken care of by human inspectors to make sure the product quality is kept in the required specifications. Since contaminations are not perfectly eliminated by machine inspections, human inspections are required to cover the un-eliminated ones. However, this is not the only reason why the human inspectors are involved in the quality control. Even though there is no contamination in the products, appearance of products may decrease the quality of the products. For example, most customers would buy Tofu with some air bubble, whereas some customers do not purchase such Tofu which contains many air bubbles, even though air bubble has nothing to do with the taste of Tofu. Therefore, human inspectors review the Tofu package to check the products containing the air bubble, which is hard to be processed by machine inspection. This study proposes image-processing-based air bubble detection method on Tofu packages to inspect the product quality. Based on the results of air bubble detection on the Tofu package, the evaluation is made on each package. The study applied an evaluation criterion based on the experiments. However, the results are not always identical to those by human inspection because of the disagreement of threshold value in evaluation. This chapter presents the image-processing air bubble detection method to determine the quality of Tofu products and discusses the feasibility of this method in comparison with the human inspection results.

T. Ito (✉)
Institute of Technology and Science, The University of Tokushima,
2-1 Minami-Josanjima, Tokushima 770-8506, Japan
e-mail: tito@tokushima-u.ac.jp

S. Fukuda (ed.), *Emotional Engineering vol. 2*, DOI: 10.1007/978-1-4471-4984-2_11,   165
© Springer-Verlag London 2013

11.1 Introduction

Consumers' requirements towards high-quality products are getting higher in many areas. In addition, food production companies are strongly required to control the high level of product quality in their production lines [1]. The contaminations such as metallic or plastic, organic materials are harmful to human body and have to be eliminated completely. Therefore, machine inspections, such as X-ray [2] or fluorescence spectrum methods [3], are used to effectively eliminate these contaminations. However, the final inspections are taken care of by human inspectors to make sure the product quality is kept within the required specifications. The reason why the human operator is involved in the final stage of production is not only the fact that the inspection cannot be carried out only by machine processing, but also the consideration that the evaluation with respect to consumers' perspectives is very critical. This is based on the observation of the authors from the discussion with manufacturing people.

In the meantime, there is a strong demand for higher quality products with lower costs and shorter time-to-market to be competitive in the worldwide marketplace. Therefore, the optimization of costs, quality and time in the product development process is critical factors to be competitive in the marketplace [4]. Furthermore, customer requirements are becoming increasingly individualized and diverse. The number of items to be considered is growing larger in the latter phase of the design–production process [5]. The appropriate decision in the upper phase of design could possibly dramatically reduce this number. These circumstances are rather regarded as profit-oriented approach.

However, sometime it would be beneficial if the customers to be involved in product design in their early stage of production. For example, emotional qualities such as product's aesthetics, which are regarded as one of the critical factors in products, should be well reviewed at the early stage of design because it would contribute to this trend of customer-oriented product development [6–8].

For food production, on the other hand, customers may sometime be involved in the product design to provide a voice of customer. However, another critical involvement of customer in food manufacturing would be in the evaluation stage. Food product is evaluated by the customers, who make the decision on the product in a store whether they purchase it or not. Therefore, the involvement of customer in the evaluation phase is critical for food manufacture [9]. Therefore, the reason why human inspector covers the final stage of inspection is not only the fact that contamination is not perfectly eliminated by machine inspections, but also the fact that appearance of products sometimes may decrease the quality of products. For example, too much air bubble in Tofu often reduces the quality of products, which are sometimes regarded as defective products. Tofu is one of the products in which air bubble strongly related to product quality. Even though air bubble has nothing to do with the taste of Tofu, customers do not purchase such Tofu which contains many air bubbles. However, machine inspection is not so good at determining the quality of Tofu in terms of air bubble occurred in production. Therefore, human inspectors review the Tofu package to eliminate the defective products because of the air bubble, which is hard to be processed by machine inspection [10].

This study is to clarify how the human inspectors make decision if the Tofu package should be delivered or to be abandoned. In order to do so, an image-processing-based air bubble detection method on Tofu packages is proposed. The processing is composed of seven sub-processes including image loading, extraction of inspection area, screening of air bubble inspection image, pre-processing, cell partitioning, air bubble detection and quality judgment. After air bubble is detected on the Tofu packages, the results are used to determine whether the product is produced to be sold or defective product not to be sold. The study applied an evaluation criterion based on the experiments. However, the results were not identical to those by human inspection because of the disagreement of threshold value in evaluation. Therefore, three different types of calculation methods, which are mean value in histogram (NMH), differential value calculation (DFV) and neural network-based calculation (NNV), were proposed and compared by using sample images. Two types of partitioning methods, cell-based partitioning (CPP) and quadtree-based partitioning (QTP), were also proposed and compared by using sample images.

This chapter presents the image-processing air bubble detection method to determine the quality of Tofu products and discusses the feasibility of this method in comparison with the human inspection results.

11.2 Tofu Production and Quality Control

Tofu is one of the very popular Japanese handmade products and can be seen in many Japanese dishes. Figure 11.1 shows some of the typical Tofu dishes. One of the simple recipes is Hiyayakko, or cold Tofu, with toppings such as shavings of dried bonito or shavings of ginger with soy source. Yudofu, or boiled Tofu, is also a very popular recipe, including blocks of Tofu simmered in hot water along with vegetables.

Traditionally, Tofu is made from soy beans by a handmade manufacturing process. Typical manufacturing process of Tofu is shown in Fig. 11.2. After soaking soy beans, the beans are grinded to crush, the crushed beans are boiled and squeezed to obtain soy milk by filtering out Okara. Okara itself is also a food for cooking. Then, adding Nigari to the soy milk, the mixtures are poured into a

Fig. 11.1 Typical Tofu recipe examples

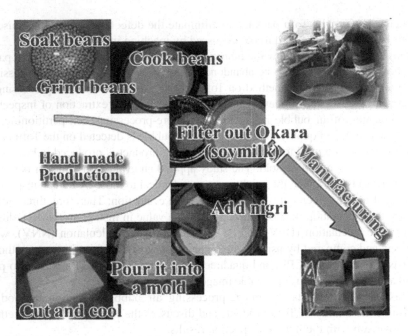

Fig. 11.2 Flow of basic Tofu production and its manufacturing production

moulding box, putting some pressure on it for a while. When Tofu is ready, it is taken out of the box, cut and cooled down. Since Tofu is perishable food, its eat-by date is very short. Moreover, Tofu is not suitable for automated production and, historically, has been one of the handmade products. However, filling machine technology [11] made it possible to implement automation of Tofu.

In the automation, soy milk and Nigari are directly filled into a plastic container and it is sealed as shown in Fig. 11.2. Then, coagulation occurs inside the sealed container, and Tofu can be made under germ-free conditions. Since the filling process is carried out under sterilization, the Tofu made by filling production has much longer product life, namely longer eat-by date. Product inspection is given to Tofu at the final stage of production to avoid contamination and damages and to provide high-quality Tofu.

Figure 11.3 shows some of the examples of contamination of foreign substances, such as hair, metal, stone, paper. Foreign substances could be checked and removed by various technical methods, for example, scanning electron microscopy (SEM), energy-dispersive spectroscopy (EDS), Fourier transform infrared spectroscopy (FTIR), vibration screening, specific gravity screening, image inspection screening, chemical analysis screening. Since contamination is not perfectly eliminated by machine inspections, human inspections are required to cover the final inspection. Even though there is no contamination of foreign substances in products, appearance of the products may increase or decrease the quality of the products. For example, if air bubble is happened in products, it often decreases

Fig. 11.3 Contamination of foreign substance examples

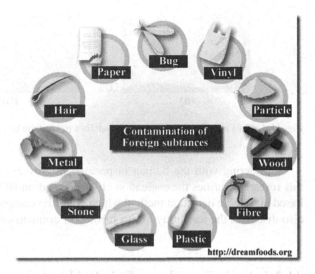

the quality of products and may become a problem in some products. Tofu is one of these products in which air bubble strongly related to product quality. Even though air bubble has nothing to do with the taste of Tofu and passes the contamination tests, some customers may not purchase such Tofu which contains many air bubbles. Therefore, human inspectors review the Tofu package to eliminate those products which contains the air bubble. These kinds of screening are hard to be processed by machine inspection methods mentioned above.

Figure 11.4a shows an example of Tofu which passed the inspection, whereas Fig. 11.4b shows an example which did not pass the inspection because of the air bubble. In both cases, taste of Tofu is similar and no harm is given to human health if it is eaten. However, some customers do not purchase the product shown in Fig. 11.4b. Therefore, the judgement is quite based on emotional feeling of customers and nothing to do with the Tofu taste or functional requirements. This emotional quality inspection is out of range of quality inspections which have been conducted so far.

11.3 Emotional Quality Inspection for Air Bubble Detection

Generally, customers see the products not only based on its functionality and its value, but also based on its aesthetic view. For example, aesthetic value is deteriorated due to air bubbles in Tofu. Aesthetic surface of Tofu is one of the product attributes that could attract emotional attachment to the customer. Air bubble in Tofu products affects consumers' feeling towards the product, which means that customers lose the feeling of purchase by air bubble. Therefore, air bubble should be avoided by any means.

Fig. 11.4 Tofu not passed the inspection and Tofu passed the inspection

Considering with the human inspectors' quality evaluation (HOV) procedures, this research clarifies the evaluation criteria based on HOV and proposes an image-based air bubble detection method which could be comparable to HOV. This section also discusses the feasibility of this method in comparison with HOV results.

11.3.1 Basic Procedure of Air Bubble Inspection

An air bubble detection procedure is designed based on a general image-processing/recognition procedure as shown in Fig. 11.5.

Process flow of air bubble detection is as follows:

11.3.1.1 Pre-processing

1. Image loading and extraction of inspection area from the loaded image.
2. Grey scaling from the loaded colour image.
3. Image histogram preparation based on the grey-scaled image.
4. Measurement of width W and maximum peak MP in the histogram.
5. Calculation of threshold value (TV) based on (W, MP) for binary image processing. (*)

11.3.1.2 Feature Extraction

6. Cell partitioning: Image is partitioned to cells CL = {cl1, cl2, ..., cnMN} by meshing M × M. (**)
7. Cell values setting: Average cell value ACV = {acv1, acv2, ... acvMN} calculation for each cell.
8. Cell binarization: ACV value is given to each cell.
9. Image binarization: Binary image processing. Based on ACV, each cell is given 0 if acv < TV, or 1 if acv > TV.
10. Block labelling: Adjacent cells are grouped together as a block cell, and each block of cells is given a unique label name BL = {bl1, bl2, ..., blN}.

Fig. 11.5 Basic flow of air bubble detection based on image-processing approach

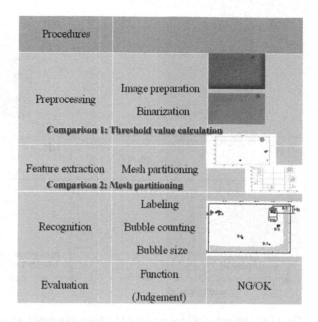

Procedures		
Preprocessing	Image preparation	
	Binarization	
Comparison 1: Threshold value calculation		
Feature extraction	Mesh partitioning	
Comparison 2: Mesh partitioning		
Recognition	Labeling	
	Bubble counting	
	Bubble size	
Evaluation	Function (Judgement)	NG/OK

11.3.1.3 Recognition

11. Centre position of each block is calculated as $BLC = \{bl1(x, y), bl2(x, y), \ldots, blN(x, y)\}$.
12. Area size of each BL is calculated as $BLS = \{bls1, bls2, \ldots, blsN\}$.

11.3.1.4 Judgment

13. Evaluation: $E = f(BL, BLC, BLS)$ is calculated to evaluate the image.

11.3.2 Histogram-Based Screening Procedure

Histogram-based screening method was used to make decision on the existence of air bubble in the image. Figure 11.6a shows a sample image with air bubble, and Fig. 11.6b shows a sample image without air bubble. Figure 11.6c, d shows the histogram for Fig. 11.6a, b, respectively. The existence of air bubble is recognized by the peak value, and the number of peaks in histogram is shown in these figures. A histogram-based screening program was implemented in this study.

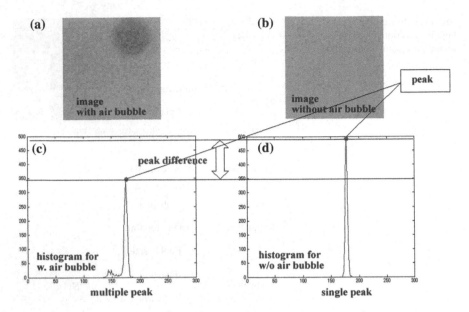

Fig. 11.6 Difference of histogram used for histogram-based screening procedure

11.3.3 Evaluation Function

As for judgment on the image to check whether it is OK or NG, the scheme
E = f(BL, BLC, BLS) is defined as the evaluation function. The evaluated values
S2–S5 are shown in Table 11.1.

Counting the number of air bubble BL, and measuring the size of air bubble
BLS, the product quality is determined as OK or NG based on the Table 11.1.

11.4 Comparison of Threshold Value Calculation

The basic procedure for air bubble detection was implemented as described in the
previous Sect. 11.3. Comparison of air bubble detection performance between
human inspector and machine inspection could be possible using this basic

Table 11.1 Evaluation criteria

	Bubble size	Allowed #
S5	>5 mm	0
S4	<5 mm, >4 mm	1
S3	<4 mm, >3 mm	3
S2	<3 mm, >2 mm	9

Note 1 * S4 is equal to 2 * S3

procedure. However, in order to improve the performance of this basic procedure, two types of modification were considered. One is threshold value calculation and the other one is partitioning procedure modification, which are shown in Fig. 11.5. This section describes the first modification, and the next section covers the second modification.

Three types of calculation method were designed and implemented in the procedure, which are mean value calculation (MNH), DFV and NNV. These values were compared with the value given by a human inspector, which is the target value (HOV). Further details regarding these calculations are shown in the following subsections.

11.4.1 Mean Value in Histogram

Mean value in the grey histogram MNH is calculated as $MNH = mean(mean(image(:, :))$. The threshold value (TV) is determined by $TV = MNH + offset$. $Offset = 0.01 * MNH$.

11.4.2 Differential Value Calculation in Histogram

Differential value is calculated around the bottom area of the peak signal using the following algorithm.

```
t = 50;
for i = 1:size(h,2) − 1
    dh = h(i + 1) − h(i);
    if(dh >= t)
        th = i;
        break;
    end
end
DFV = th −1;
```

11.4.3 Neural Network-Based Calculation

Peak value and bottom width of the envelope curve were read from the density histogram as an input signal for neural network. The target threshold is determined by a human inspector as a training datum. Then, the threshold value is calculated by the trained neural network. Table 11.2 shows an excerption from the training data as an example.

Table 11.2 Excerption of neural network training data

P	#	1	2	3	4	5	6	7	8	9	10	11	12
	PK	183	187	178	171	161	171	155	161	194	200	180	192
	WD	29	3	30	20	22	28	38	28	28	29	22	18
T	TV	180	183	175	166	156	164	153	158	192	195	176	189

11.4.4 Comparison of Three Methods of Calculation

Threshold value (TV) in 72 samples was calculated by the three methods described above. These were compared each other to see if there is any significant difference as opposed to HOV. The subsections below show the results of comparison in terms of TV itself, bubble detection and quality evaluation using the evaluation function.

11.4.4.1 Comparison of TV

Figure 11.7 shows the result of comparison regarding TV in each sample. Significant difference was recognized between the TV values calculated by three methods and HOV, in sample #54, #71, #72 and #83. Overall, NNV is the most close to HOV.

11.4.4.2 Comparison of Bubble Detection

Figure 11.8 shows the result of comparison regarding bubble areas in each sample. Overall, NNV is most similar to HOV.

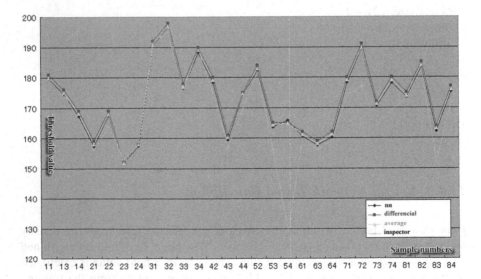

Fig. 11.7 Comparison of TV value by MNH, DFV and NNV

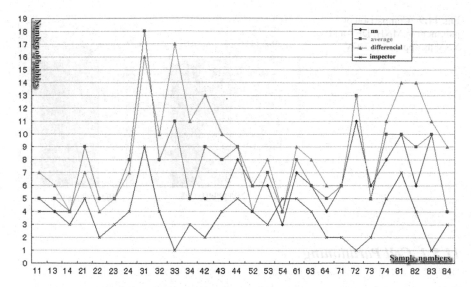

Fig. 11.8 Comparison of bubble area detection by MNH, DFV and NNV

Figure 11.7 shows the graph of error in MNH, DFV and NNV from the target value HOV. Bigger error was observed in MNH and DFV than that of NNV.

11.4.4.3 Comparison of Quality Evaluation

Using the evaluation function E, the sample images were evaluated based on the parameters in three calculation methods. This experiment used 32 image samples including 10 NG samples and 22 OK samples. Figure 11.9 shows the results of judgment on the NG samples on the left side of the graph and the OK samples on the right half of the graph. There was no significant difference in these results.

11.5 Comparison of Partitioning Method

This section covers another modification in the basic procedure, which is the modification in partitioning procedure. Two types of partitioning were proposed and studied in this research. Image processing could be performed without partitioning the image, which is pixel-based image recognition. This research is to propose an image-based bubble recognition, which could be applied in product inspection in the final stage of production. Therefore, the shorter process time is required. Even if the accuracy of image recognition is lower than that of pixel-based approach, performance of process is critical. Therefore, image area partitioning in image recognition process is considered in this research. Two types of partitioning are under study, which are CPP and QTP. The following two subsections cover these two partitioning methods.

Fig. 11.9 Comparison of judgment in MNH, DFV and NNV as opposed to HOV

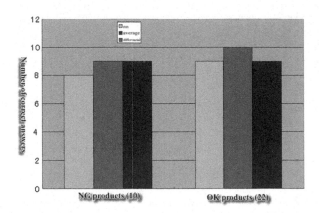

11.5.1 Cell Partitioning

The cell partitioning presented in this subsection is the default method in the basic procedure. In other words, the comparisons in the previous section adopted the CPP in the experiment. In cell partitioning, the image to be processed is partitioned by the specified number of cells; then, the whole image is given binary process. Figure 11.10 shows the results of 60 cells CPP, which reduces the process time as opposed to the pixel-based processing. Even though the accuracy of air bubble shape could be deteriorated according to the number of partitioning cells, the contour of the air bubble could be kept if the appropriate number is applied.

11.5.2 Quadtree Partitioning

Quadtree partitioning is another approach for studying to improve the performance of air bubble recognition in the image. The image under process is checked by the histogram-based screening, and the image is partitioned into four areas if any bubble is detected. Then, the same procedure is given to each partitioned area as long as any bubble is detected by the screening process. Figure 11.11 shows the results of image partitioning by the QTP procedure.

Figure 11.12 shows the original image shown in a and its results of air bubble recognition by CPP and QTP, as shown in b and c, respectively.

Table 11.2 shows an example of the results of judgment by CPP, QTP and HOV, of which image was judged as NG in all cases.

As a result of evaluation on 72 image samples, correct judgment of CPP is 62 out of 72 images (86.1 %), whereas that of QTP is 61 out of 72 images (84.7 %). Therefore, no significant difference was recognized between these two methods (Table 11.3).

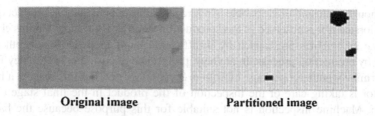

Original image   **Partitioned image**

Fig. 11.10 Original image versus partitioned image

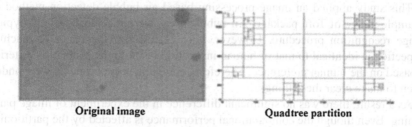

Original image   **Quadtree partition**

Fig. 11.11 Original image versus quadtree-partitioned image

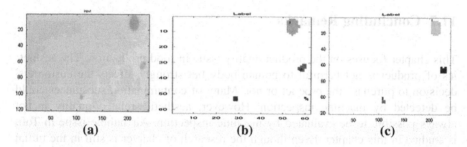

(a)   (b)   (c)

Fig. 11.12 Original image, bubble recognition by CPP and bubble recognition by QTP

Table 11.3 Comparison of judgment by cell partitioning, quadtree partitioning and human inspection

	2	>2; <3	>3; <4	>4; <5	>5	Total	Judge
CPP	2	1	0	0	1	4	NG
QPT	0	1	1		0	1	3	NG
HOV	1	1	1	0	1	4	NG

11.6 Results and Discussions

Product quality is very important in any field of production. In addition, food production companies are paying much more focus on product quality. Contamination should be avoided by all means because it is very harmful to human body.

Even though inclusion of air bubble is not regarded as contamination, product quality is deteriorated if it is observed in food production, such as Tofu studied in this chapter.

Some air bubbles are caused by insufficient coagulation. However, others are caused by production process. In anyhow, preference of customers is the key factors to determine whether a package of Tofu is accepted or not. For this reason, a human inspector is taking care of the inspection of the product in the final stage of production. Machine inspection is not suitable for this purpose because the basis of evaluation is regarding human preference. In addition to the handmade production of Tofu, machine-made production is getting popular in Tofu. Therefore, machine inspection is under consideration to screening out Tofu with air bubble.

This study applied an image-processing-based air bubble detection method on a sample image of Tofu package. Air bubble areas were recognized by the typical image recognition procedure. However, it is not clear if the result of machine inspection is identical to that of human inspection because the evaluation criterion is based on the human preference. Therefore, comparison experiments were undertaken to make clear these things.

As a result, there was no significant difference in the experiment of image partitioning. Even though the computational performance is affected by the partitioning methods, it was not related to the results of evaluation.

11.7 Concluding Remarks

This chapter focuses on the product quality issue in food production. The aesthetics of product is not harmful to human body but strongly affects the customers' decision to purchase the product or not. Many of contamination substances could be detected by machine inspection. However, aesthetics-related quality is not always possible to be evaluated by machine inspection. Air bubble issue in Tofu is studied in this chapter. Even though the research of chapter is still in the initial stage, this research took a small one step to clarify how the human inspector determines the decision if the product is good or bad. Further study will be continued using video camera streaming data as well as a large number of digital image data.

References

1. Tokyo Marine Risk Consulting Co. (2001) Current situation and counter measures to foreign substances in foods, risk radar, pp 1–8 (in Japanese)
2. Ohira N, Hongjun Z, Sakamaki K, Kamimura K, Shimizu H, Saiki H (2008) X-ray inspection of foreign bodies in foods. Bull TIRI 3:18–21 (in Japanese)
3. Kato A, Sasaoka H, Akida D (2005) Development of detection technology for non-metallic foreign substances in foods. Savemation Rev 2:48–55 (in Japanese)
4. Prasad B (1996) Concurrent engineering fundamentals, integrated product and process organization, vol 1. Prentice Hall, New Jersey

5. Ito T, Fukuda S (1995) A methodology of perspective browsing to support concurrent design. In: Proceedings of the 1995 ASME design engineering technical conference, vol 2. Boston, pp 687–693
6. Shuichi Fukuda (2011), Emotion and process quality. In: Proceedings of DECT'01/ASME international 31st computers and information in engineering (CD-ROM, DETC2011-48663)
7. Widiyati K, Aoyama H (2011) A study of kansei engineering in PET bottle silhouette. In: Proceedings of DECT'01/ASME international 31 st computers and information in engineering (CD-ROM, DETC2011-48066)
8. Yanagisawa H, Murakami T (2007) Emotional shape generation system with exchange of other's viewpoints for externalizing customers' latent sensitivity. In: Proceedings of DECT'07/ASME international 27th computers and information in engineering (CD-ROM, DETC2007-34726)
9. Maslow AH (1943) A theory of human motivation. Psychol Rev 50:370–396
10. Kano N, Seraku N, Takahashi F, Tsuji S (1984) Attractive quality and must-be quality for quality. J Jpn Soc Qual, Control, pp 39–48
11. Kawamura S (1991) Device for preventing liquid from dripping from filling nozzle of liquid filling machine. US patent US 5,016,687, 21 May 1991 (Inventor)

Chapter 12
Design Impression Analysis Based on Positioning Coloring of Design Elements

Hideki Aoyama and Naoki Okamura

Abstract Since it is becoming increasingly difficult to differentiate products by function and quality as manufacturing technology continues to improve, there are strong demands for product designs that reflect customer preferences. In this study, method for analyzing design impression based on the positioning and coloring of design elements is proposed. The methods are applied to the design of a portable music player as the object for analysis to verify the usefulness of the method.

12.1 Introduction

Since it is becoming increasingly difficult to differentiate products by function and quality as manufacturing technology continues to improve, there are strong demands for product designs that reflect customer preferences at an early stage of product development [1–3]. Moreover, it is also important to analyze impressions that consumers can have and evaluate the adaptability of design ideas and consumer preferences quantitatively and objectivity.

For this reason, numerous researches proposing impression analysis techniques of product designs have been carried out. Sekiguchi proposed design evaluation structure analysis using multivariate analysis and rough sets [4]. With this method, design shapes such as body shape and roundness of the corner are set as recognition features of products for evaluating design impression. Hung Cheng proposed an automatic design support system and evaluated impression of two-colored products using genetic algorithm based on the relation between two colors [5]. Most of such conventional methods, however, are based on only one shape or coloring, and the objects for evaluation are limited.

H. Aoyama (✉) · N. Okamura
Department of System Design Engineering, Keio University, 3-14-1 Hiyoshi,
Kohoku-ku, Yokohama 223-8522, Japan
e-mail: haoyama@sd.keio.ac.jp

N. Okamura
e-mail: okamura@ina.sd.keio.ac.jp

S. Fukuda (ed.), *Emotional Engineering vol. 2*, DOI: 10.1007/978-1-4471-4984-2_12,
© Springer-Verlag London 2013

In this study, to support the designer's activities in the product design process, methods to analyze design impressions based on positioning and coloring of design elements are proposed and are applied to the design of a portable music player as the object for analysis to verify the usefulness of the methods.

12.2 Proposal of Design Impression Analysis Methods

12.2.1 Outline of Proposed Methods

The positioning and coloring of design elements are treated as independent evaluation parameters, and the proposed methods are based on these viewpoints. Design elements are the individual parts making up a product. For example, in the case of the digital camera as shown in Fig. 12.1, the lens, flash, and logo mark are the design elements.

12.2.2 Design Impression Analysis Method Based on Positioning of Design Elements

Design impression changes greatly depending on how design elements are positioned [6]. Given the countless number of combinations of the positioning of design elements, it is difficult to analyze impression for all combinations. In conventional methods, this problem is handled by setting up some levels in design factors. Therefore, the object of analysis changes simply and limitedly.

The following techniques are therefore proposed. First, the positioning of design elements is quantified by defining some original parameters. These parameters are called positioning quantify parameters (PQPs). Secondly, the positioning evaluation functions (PEFs) are defined between PQPs and impression evaluation values which indicate the degree of consumer's impression. Finally, to analyze design impression based on positioning, the impression analysis value of positioning is calculated by using functions. Since all the designs can be quantified and analyzed by this method, it becomes possible to achieve the expansion of object for analysis.

1. **Definition of Positioning Quantify Parameters**

(a) Gravity E_g: E_g is defined from the y coordinates of the gravity point of design element i (y_i) and area of design element i (S_i) in Eq. (12.1), and its range is [0, 1]. When the E_g is big, it is understandable that the gravity point of the whole design must be high. On the other hand, the smaller E_g means that the gravity point must be low. Consequently, the larger E_g gives the more stable impression, and the smaller E_g represents the more unstable impression.

Fig. 12.1 Example of design elements

(b) Symmetry E_s: When considering a design (A) and a design (A') which is the reverse of design (A) at the y-axis, S_c is defined as an overlapping area between design (A) and design (A'). E_s is defined from S_c and area of design element i (S_i) in Eq. (12.2), and its range is [0, 1]. The larger E_s is, the more symmetrical will the positioning pattern be, and the smaller E_s is, the less symmetrical will the positioning pattern be. Besides, the larger E_s is, the more stable will design impression be. The smaller E_s is, the more unstable will

$$E_g = \sum_{i=1}^{n} (y_i \times S_i)$$
(12.1)

design impression be.

$$E_s = \frac{S_c}{\sum_{i=1}^{n} S_i}$$
(12.2)

(c) Direction E_d: E_d is defined from the inclination of the straight line when the gravity point of design element i (x_i, y_i) is approximated as a linear equation in Eq. (12.3), and its range is [0, 1]. The larger E_d is, the more will the direction of positioning lean by 45°. The smaller E_d is, the more will the direction of positioning become vertical or horizontal. Besides, the larger E_d is, the more stable will design impression be. The smaller E_d is, the more unstable will design impression be.

$$E_d = \frac{n \sum_{k=1}^{n} x_k y_k - \sum_{k=1}^{n} x_k \sum_{k=1}^{n} y_k}{n \sum_{k=1}^{n} x_k^2 - \left(\sum_{k=1}^{n} x_k\right)^2}$$
(12.3)

2. Definition of Positioning Evaluation Functions

Three positioning evaluation functions: PEFs, f_g (E_g, j), f_s (E_s, j), f_d (E_d, j), were defined between E_g, E_s, E_d, and impression evaluation value μ_j of KANSEI-word j. In this study, KANSEI-word means an adjective describing design impression.

Basic experiment was carried out to define the PEFs and acquire KANSEI-data $[E_g, E_s, E_d, \mu_j]$. The functions were defined by plotting this data in the rectangular coordinate system where the x-axis represents the quantification positioning

Fig. 12.2 Example of definition of PEF about *KANSEI*-word "powerful"

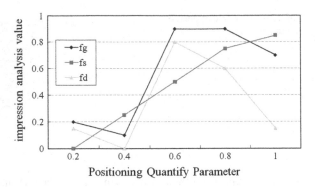

parameter and *y*-axis the impression evaluation value, and implementing linear interpolation between data. An example of definition is shown in Fig. 12.2.

3. **Calculation of positioning impression analysis values for analysis based on positioning of design elements**

When a design whose PQPs are e_g., e_s, e_d is entered, the impression analysis value μ_{pos} (*j*) based on the positioning of the KANSEI-word *j* can be calculated by Eq. (12.4).

$$\mu_{pos}(j) = \frac{f_g\left(e_g, j\right) + f_s\left(e_s, j\right) + f_s\left(e_s, j\right)}{3} \tag{12.4}$$

From μ_{pos} (*j*), design impression can be analyzed based on the positioning of design elements.

12.2.3 Design Impression Analysis Method Based on Coloring of Design Elements

Design impression changes greatly depending on how the coloring of design elements is positioned [7–13]. When deciding coloring of design activity, the PCCS color system (Practical Color Coordinate System) is often used. With this system, the chroma and value of Munsell systems are unified into Tone which represents color condition, and the color is expressed by Hue and Tone. When expressing images using the PCCS color system, the following procedure is generally used.

First, to determine the outline of the image, the image of the hue or tone as shown in Fig. 12.3 is examined closely, and a color suitable for the image is chosen as the representative color. Next, to fine-tune the image, other colors are decided while adjusting the hue difference or tone difference to the representative color. In this study, if the representative color is chosen from the image of hue, this coloring pattern is called "H-type coloring". On the other hand, if the representative color is chosen from the image of tone, this coloring pattern is called "T-type coloring".

Fig. 12.3 Color model of PCCS and impression of color

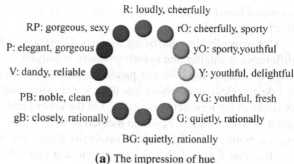

R: loudly, cheerfully

RP: gorgeous, sexy

P: elegant, gorgeous

V: dandy, reliable

PB: noble, clean

gB: closely, rationally

rO: cheerfully, sporty

yO: sporty, youthful

Y: youthful, delightful

YG: youthful, fresh

G: quietly, rationally

BG: quietly, rationally

(a) The impression of hue

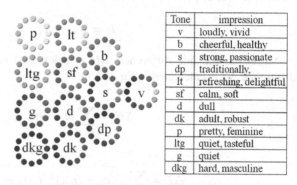

Tone	impression
v	loudly, vivid
b	cheerful, healthy
s	strong, passionate
dp	traditionally,
lt	refreshing, delightful
sf	calm, soft
d	dull
dk	adult, robust
p	pretty, feminine
ltg	quiet, tasteful
g	quiet
dkg	hard, masculine

(b) The impression of tone

This study proposes the following analysis algorithm based on this procedure. First, the design element best conveying impression is picked from the group of design elements, and the color of this design element is decided as the representative color. Next, the impression evaluation value of the representative color is derived from the neural network.

Finally, the impression evaluation value of the representative color is corrected according to the hue difference or tone difference between the representative color and other colors. The value after correction is considered to be the impression analysis value of coloring, and from this value, design impression based on coloring is analyzed.

By using the neural network [14], it becomes possible to analyze impression of all colors which are in HCV color system. Moreover, by accepting coloring technique to method, relative analysis between the colors becomes possible, and it is then realized to achieve the expansion of object without limiting to number/combination of colors.

1. **Deciding the representative color**

Generally, to choose a color which easily conveys impression when fine-tuning the image to be expressed, if the coloring pattern is H-type, the hue of other colors is arranged based on the hue of the representative color, and the image is fine-tuned by changing the tone. If the coloring pattern is T-type, the tone of other colors is arranged based on the tone of the representative color, and the image is fine-tuned by changing the hue.

Consequently, in comparing the hue difference and tone difference, if the hue difference is smaller, the coloring pattern is judged as H-type, or if the tone difference is smaller, the coloring pattern is judged as T-type. Moreover, with H-type, the design element m whose hue is h_m is the median of the hue h_i ($i = 1, 2,..., n$) is decided as the representative color. On the other hand, with T-type, the design element m whose tone is T_m is the median of tone T_i ($i = 1, 2,..., n$) is decided as the representative color. The tone is originally a concept, and it does not have a value. So, the tone T_i ($i = 1, 2,...,n$) was calculated from chroma c_i ($i = 1, 2, ..., n$) and value v_i ($i = 1, 2,..., n$) in Eq. (12.5).

$$T_i = \sqrt{c_i^2 + v_i^2} \tag{12.5}$$

Moreover, the hue difference ΔH is calculated in Eq. (12.6), and the tone difference ΔT is calculated in Eq. (12.7).

$$\Delta H = \sum_{i=1}^{n-1} |h_i - h_{i+1}| \tag{12.6}$$

$$\Delta T = \sum_{i=1}^{n-1} \sqrt{(c_i - c_{i+1})^2 + (v_i - v_{i+1})^2} \tag{12.7}$$

2. Deriving the impression analysis value of the representative color using the neural network

The single color image scale is known for the color impression of KANSEI-word [8]. However, the concrete composition method of this scale is not clear, and the scale corresponds only to limited colors.

Therefore, in this study, single color impression by using a neural network was utilized to analyze the impression of a color. The input layer has three units: hue H, chroma C, and brightness V, while the output layer has a unit impression analysis value of a single color. The color impression is investigated, and the results of the investigation are studied by back-propagation, and a hierarchical neural network that derives single color impression value from HCV value is built. From this network, the impression analysis value of the representative color v_j of KANSEI-word j is derived from the HCV value of the representative color.

3. Calculating the impression analysis value of coloring by correction functions

The correction functions $f_H(\Delta H, j)$, $f_T(\Delta T, j)$ are defined between the hue difference ΔH, tone difference ΔT, and correction degree Δv_j of the impression evaluation value of representative color v_j. Basic experiments to define correction functions were conducted, and KANSEI-data [ΔH, ΔT, Δv_j] were acquired. The

Fig. 12.4 Example of
definition of correction
function about *KANSEI*-word
"powerful"

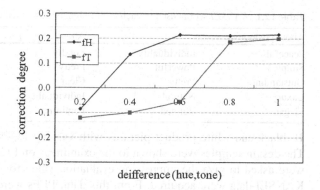

deifference (hue,tone)

correction functions are defined by setting this data in the rectangular coordinate
system where the *x*-axis is ΔH and ΔT, and the *y*-axis is the correction degree
Δv_j, and implementing linear interpolation between the data. An example of definition is shown in Fig. 12.4.

With H-type, the image to be expressed is fine-tuned by the tone difference. For
this reason, the impression analysis of coloring value $\mu_{col}(j)$ of KANSEI-word *j* is
calculated from the correction function $f_T(\Delta T, j)$ in Eq. (12.8).

$$\mu_{col}(j) = v_j + f_T(\Delta T, j) \tag{12.8}$$

With T-type, the image to be expressed is fine-tuned by the hue difference. So,
the impression analysis of coloring value $\mu_{col}(j)$ of KANSEI-word *j* is calculated
from the correction function $f_H(\Delta T, j)$ in Eq. (12.9).

$$\mu_{col}(j) = v_j + f_H(\Delta H, j) \tag{12.9}$$

12.3 Example of Applying Proposal Method

The analysis of the design impression based on the proposed methods for the
design of a portable music player was attempted, and the following basic experiments (1) and (2) needed for each analysis method were conducted. The examinees were 13 young parsons (eight males and five females). The design of a
portable music player is composed of a simple figure to exclude the influence of
the shape not considered in the proposed method.

12.3.1 Experiment (1)

Experiment (1) was conducted to acquire KANSEI-data for defining PEFs. The
KANSEI-words, shown in Table 12.1, consisted of 10 words chosen by cluster
analysis which express impression based on positioning. The design samples were

Table 12.1 *KANSEI*-words used in each impression analysis method

About positioning		About coloring	
Peaceful	Individuality	Loudly	Elegant
Simple	Powerful	Pretty	Polished
Outstanding	Pretty	Clear	Reliable
Luxury	Natural	Individuality	Familiar
Elegant	Refined	Feminine	Composed

18 black-and-white product pictures with varying PQPs as shown in Fig. 12.5a. The design samples were shown to the examinees on LCD monitor, the examinees were asked to give the impression evaluation value for each KANSEI-word, and KANSEI-data were acquired. From this data, PEFs were defined. The impression evaluation values were given 11 stages in the range of [0, 1].

12.3.2 Experiment (2)

Experiment (2) was conducted for acquiring KANSEI-data for defining the correction functions. For the KANSEI-words, as shown in Table 12.1, 10 words were chosen by cluster analysis which express impression based on coloring. The design samples were 14 two-colored pictures composed of standard colors ($H = 0.0$, $C = 0.9$, $V = 0.9$) and other colors with varying hue difference or tone difference with standard color as shown in Fig. 12.5b. The design samples were shown to the examinees on LCD monitor, the examinees were asked to give the impression evaluation value for each KANSEI-word, the difference between the impression analysis value of standard color and that of two-colored images was regarded as correction value, and KANSEI-data were acquired. From this data, the correction functions were defined. The impression evaluation values were given 11 stages in the range of [0, 1].

Each analysis method was implemented based on the functions defined from each basic experiment. Impression analysis for the design shown in Fig. 12.6a based on positioning and coloring using the implemented method was attempted. Figure 12.6b shows the results of impression analysis from positioning, and Fig. 12.6c shows the result of impression analysis from coloring.

12.4 Experiments to Assess Proposed Method

To confirm the usefulness of each method, assessment experiments to compare the analysis results of the implemented methods with the answers of the examinees were conducted. The proposal technique constructed with Chap. 3 was mounted as an analysis system, and impression of the design samples newly made was analyzed by this system. And the examinees answered the impression of the samples. Comparison between the result of system and the result of examinee's answer was made, and whether the proposed method can analyze in conformity with human's sensibility was examined.

12.4.1 Experiment Details

The following assessment experiments (1) and (2) were executed. The examinees consisted of 13 young parsons (eight males and five females) different from the examinees of basic experiments.

1. Experiment (1)

Assessment experiments on impression analysis based on positioning: Design samples consisting of 18 black-and-white product pictures with varying PQPs were shown to the examinees on a LCD monitor, and the examinees were asked to give the impression evaluation value for each KANSEI-word used in basic experiment (1). The design samples were also analyzed using the proposed analysis method, and results of analysis based on positioning were acquired.

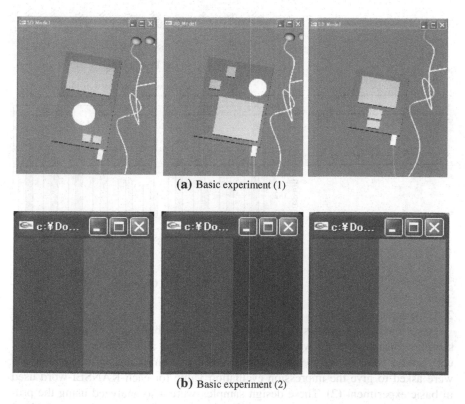

(a) Basic experiment (1)

(b) Basic experiment (2)

Fig. 12.5 Example of sample for basic experiments

Fig. 12.6 Results of design impression analysis by proposed methods

(b) Analysis based on position

(a) Sample design

(c) Analysis based on coloring

Fig. 12.7 Results of assessment experiment

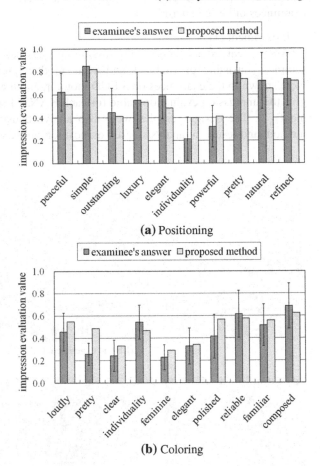

■ examinee's answer　□ proposed method

impression evaluation value

peaceful, simple, outstanding, luxury, elegant, individuality, powerful, pretty, natural, refined

(a) Positioning

■ examinee's answer　□ proposed method

impression evaluation value

loudly, pretty, clear, individuality, feminine, elegant, polished, reliable, familiar, composed

(b) Coloring

2. Experiment (2)

Assessment experiments on impression analysis based on coloring: Design samples consisting of 12 two-colored product pictures and 12 three-colored product pictures were shown to the examinees on a LCD monitor, and the examinees were asked to give the impression evaluation value for each KANSEI-word used in basic experiment (2). These design samples were also analyzed using the proposed analysis method, and results of analysis based on coloring were acquired.

Table 12.2   Results of t-tests

(a) Positioning

	Sample 1		...	Sample 15	
	Examinee's answer	Proposed method	...	Examinee's answer	Proposed method
Peaceful	3	8		5	7
Simple	2	3		4	3
Outstanding	8	9		8	8
Luxury	6	7		7	5
Elegant	5	6		6	6
Individuality	10	10	...	10	10
Powerful	9	5		9	9
Pretty	7	4		2	1
Natural	1	1		1	4
Refined	4	2		3	2
ρ-value		0.648			0.878
t-value		2.41	...		5.20
Significance level 5 %		○			○

(b) Coloring

	Sample 1		...	Sample 24	
	Examinee's answer	Proposed method	...	Examinee's answer	Proposed method
Loudly	2	3		3	10
Pretty	10	9		8	6
Clear	7	5		1	8
Individuality	1	7		6	7
Feminine	9	10	...	5	9
Elegant	5	6		6	5
Polished	6	4		2	4
Reliable	2	2		8	2
Familiar	8	8		10	3
Composed	4	1		3	1
ρ-value		0.655	...		−0.291
t-value		2.45			−0.86
Significance level 5 %		○			×

12.4.2   Result of Assessment Experiments

The results of assessment experiment (1) are shown in Fig. 12.7a, and those of assessment experiment (2) are shown in Fig. 12.7b. The average margin of error between the answers method is 0.12, suggesting that the proposed methods were able to analyze design impressions closely to human sensitivity.

To confirm the usefulness of each method proposed, significance was verified using the rank correlation coefficient ρ. The rank correlation coefficient ρ is an index which represents correlation between two data by comparing two ranking data. In this study, based on rank data where the impression analysis values of ten KANSEI-words were ranked in ascending order, the correlation between the answers of the examinees and the results of the proposed method was evaluated using the rank correlation coefficient ρ.

The results of rank correlation coefficient calculation and t-tests are shown in Table 12.2. The t-tests on the usefulness of the rank correlation coefficient confirmed its significance in 12 out of the 15 samples used in assessment experiment (1) as shown in Table 12.2a as well as the significance of the proposed method based on positioning. The significance of the rank correlation coefficient was also confirmed in 15 out of 24 samples used in assessment experiment (2) as shown in Table 12.2b as well as the significance of the proposed method based on coloring.

12.5 Conclusion

In this study, the following design analysis methods based on the positioning and coloring of design elements were proposed.

(a) The PQPs were defined, and analysis based on positioning was enabled by defining the PEFs between PQPs and KANSEI-words.
(b) A representative color was selected from design elements, and the impression analysis value of the representative color was derived from the neural network. Analysis based on coloring was enabled by correcting the impression analysis value of the representative color from the hue difference and tone difference.

The proposed methods were also applied to the design of a portable music player, and verification of its usefulness through assessment experiments indicated that it enables designers to analyze design impressions quantitatively and objectively.

By integrating each analysis method, there is a possibility that the design candidate creation system for supporting designers will be constructed. For example, when designer has a demand that he/she would like to change the impression of initial design without changing colors, creation system can propose the design candidate suitable for the demanded impression by changing positioning of initial design. If the analysis of initial design and the creation of design candidate are done consistently, it is expected to contribute to the designer's design activity greatly.

References

1. Hirokawa M (2000) Analysis of relation between design evaluation words and form elements. Japan Soc Sci Des 47:134–135
2. Lai H-H (2006) User-oriented design for the optimal combination on product design. Int J Prod Econ 100(2):253–267

3. Zhai L-Y (2009) A rough set based decision support approach to improving consumer affective satisfaction in product design. Int J Ergon 39(2):295–302
4. Sekiguti A (2007) The design evaluation structure analysis of the portable audio product by multivariate analysis and rough sets. Japan Soc Sci Des 54:49–58
5. Cheng H-Y (2007) Automatic design support and image evaluation of two-colored products using color association and color harmony scales and genetic algorithm. Comput Aided Design 39(9):818–828
6. Mitui H (2006) Shin Kouseigaku. Rokkisya, pp 12–29
7. Shiranita K (2004) An impression-word-giving system for images based on colors. Japan Soc Mech Eng 70(689):192–199
8. Kobayashi S (2001) Color Image Scale, vol 22. Nippon Color and Design Research Institute, pp 6–11
9. Cooper E, Kamei K (1999) Development of a color balance support system. Trans Hum Interface Soc 1(4):73–80
10. Hsiao S-W (2006) An image evaluation approach for parameter-based product form and color design. Comput Aided Des 38(2):157–171
11. Ou L-C (2004) A study of color emotion and color preference. Part 1: Color emotions for single colors. Color Res Appl 29(3):232–240
12. Ronnier M (2006) Applying color science in color design. Opt Laser Technol 38:392–398
13. Wu F-G (2010) Effects of color sample display and color sample grouping on screen layout usability for customized product color selection. Comput Hum Behav 98(2):51–60
14. Ishihara S (1995) An automatic builder for a KANSEI expert system using self-organizing neural networks. Int J Ind Ergon 15(1):13–24

Chapter 13
Robust Design on Emotion for PET Bottle Shape Using Taguchi Method

Khusnun Widiyati and Hideki Aoyama

Abstract This study evaluates the relationships between PET bottle shapes and customers impression using kansei engineering. The objective of the study is to understand the design parameter that can satisfy customers' impression. Robust design of Taguchi method was applied to obtain the optimum combination of design parameters that can match customers' impressions. The optimum combination of design parameters can generates a PET bottle design that stable to the inconsistency of customers' impressions. Factor analysis and neural network were used to map the relationship between shape impressions to design parameters. Kano model was integrated to the proposed method to enhance the PET bottle design so that the design can satisfy the customers' impressions. The proposed robust design method was implemented in Solidworks' API to create user interactive system. The system can helps designer to design a personal PET bottle shape based on the desired impressions.

13.1 Introduction

In the early concept design, most of designer's attention is paid on product function rather than product appearance. It can be observed clearly on the products developed decades ago which was mostly were in a big sizes. Todays, the way things look and how they make customer feel is very important in customer choices.

Designer of customer product now target on reaching the feeling experiences for their customers, with some even looking on the emotion explicitly as the starting point for design. It is how the product design can affect customer's emotion. It is known as aesthetic design and it is the way how product connects with customers emotion needs in aspiration with emotion experiences. When a product can satisfy customers' emotion, the product will give deep impact to the owner, and as a

K. Widiyati (✉) · H. Aoyama
Department of System Design Engineering, Keio University, 3-14-1 Hiyoshi,
Kohoku-ku, Yokohama 223-8522, Japan
e-mail: khusnun@ina.sd.keio.ac.jp

H. Aoyama
e-mail: haoyama@sd.keio.ac.jp

result, it will be treasured by the owner. Using this perspective, industries are shifting their way for increasing the aesthetic value which can give benefits not only to customers, but also to the industries themselves (profit) [1].

Let's take a look on the design of PET bottles distributed in Japan. In Japan there are many kind of PET bottle shape which can be found. At the time when beverage industries manufactured the PET bottle, it is clear that Japan citizens have a considerably high aesthetic level. Most of Japan citizens consider every aspects of their appearance, which also include daily product, as something that can defines their emotions. Observation to the design of PET bottles, it seems that designer of beverage industries in Japan decided to target the customers' desire, which is to create an eye catching package that close to customers' emotions. It is expected that customers could obtained difference emotion experiences by using a unique bottle shapes offered by the industries. Regardless to what kind of product image attempted to be embedded in the PET bottle shape by beverage industries, nevertheless, there is no standard rule-of-thumb which regulates the design of PET bottle shape and the associated impressions evoked. In our research, we are interested to understand the relation between the uniquely manufactured bottles shape with customers emotions.

13.1.1 Robust Design

The terms of robust design has been used very often to describe the stability of a method in different environment with various levels of noise factors. Robust design is a technological breakthrough that enables efficient experimental and simulation methods to optimize product and process for manufacturability, quality and reliability. The role of robust design is to minimize the sensitivity of products and processes to uncontrollable noise factor (e.g., environmental, deterioration and manufacturing). Robust design achieves this by selecting the objective function and evaluating it to maximize the function with regards to controllable design factors. The evaluation in robust design involves the application of orthogonal array, as an efficient analytical method for reducing the variability and adjusting to target.

The robust design started when Genichi Taguchi [2] gave new paradigm to the traditional experimental design method with his emphasis on the inclusion of noise factors. Taguchi's approach offered a method to calculate the noise factors using the signal-to-noise ratio.

The development of the robust design until it reached the current stage started as one approach to reduce a product's or process's functional variation by reducing the variability due to the noise factors or eliminate them entirely. However, this is not always feasible or even possible. Indeed, any attempt to reduce the variability caused by noise factors would mean the reduction of the useful range, that demanding a tighter manufacturing tolerance or specifying low drift parameters. All this methods quickly raise the cost of the product or process and are inherently not efficient. A better method is to center the design parameters to minimize sensitivity to noise factors through the principle of robustization.

13.1.2 Taguchi Method in Functional Design

The Taguchi method is one of the tools for achieving optimized and robust response. As stated by Taguchi that products or processes have good quality if it can perform its intended functions without variability, and causes little loss through harmful side effect, including cost of using it. From this statement, it is clear that the Taguchi method emphasizes on losses or cost, which commonly termed as the quality loss function. The quality of a product can be expressed as the total loss to society from the time the product is sold to customer. This loss is incurred largely due to the deviation of the functional characteristic from the target value and is estimated through the quadratic loss function. The loss function is used to define performance measures of a quality characteristic of a product or process.

To achieve the product or process quality design, Taguchi method offers three phases: system design, parameter design and tolerance design. In the system design, designers use their experience to create functional design. The product can be a new product or process with an improved modification of an existing product or process. The parameter design determines the optimal setting of the product or process parameters. These parameters have been identified during the system design phase. Design of parameter method is applied in this stage to determine which controllable factors and which are the noise factors are the significant variables. The aim is to set the controllable factors at those levels that will result in a product or process being robust with respect to noise factors. The Taguchi method solved the problem by minimize the variability of products and the processes for improving the quality and reliability.

13.1.3 Factors and Orthogonal Array

In the manufacturing, the system design can be simplified as a process to achieved objective function with the influences of factors. The factors are identified as signal factors, control factors and noise factors. Signal factors are set by the designer to attain the desired output. Control factors are the parameters values set by the researcher. Each of the control factors is studied at least at two-levels and the parameters design objective is to select the best level. Since there are many control factors in an experiment, the factors are represented in a matrix notation. While noise factors are not controlled by the researchers or the designer. However for the purpose of the optimization, these factors may be set at one level or more levels. Since there are many noise factors in an experiment, these factors are represented in a matrix notation. Obviously, no optimum noise factor is selected in an experiment.

It is important that factors are orthogonal to each other and that they are balanced. An orthogonal array is used as the basis for balanced comparison of several factors. The orthogonal array also simplifies the data analysis. Many orthogonal arrays are available with different factor and level combination.

13.1.4 Performance Measure

To improve the quality of both the average response of a quality and its variation, the Taguchi method proposed the single measurement using Signal-to-noise ratio. Signal to noise (S/N) ratio is used to measure the performance of each factor combination, in the form of a response or quality characteristic. Consequently, Taguchi method selects the design parameter levels that will maximize the appropriate S/N ratio. There are three categories of signal to noise ratio, smaller the better, larger the better and nominal the best. Smaller the better response is used to measure a function with ideal value of zero. The decrease in response value increases quality. Nominal the better is used to measure a function with a specific target that considered ideal. And lastly, larger the better is used to measure a function in which infinity is the ideal value. The increase in response value increases quality. Smaller the better, nominal the better and larger the better are formulated in the following equation:

Smaller the better

$$
S/_{N} \, ratio = -10 \times \log \left(\frac{1}{n} \sum_{i=1}^{n} y_i^2 \right) \tag{13.1}
$$

Larger the better

$$
S/_{N} \, ratio = -10 \times \log \left(\frac{1}{n} \sum_{i=1}^{n} \frac{1}{y_i^2} \right) \tag{13.2}
$$

Nominal the best

$$
S/_{N} \, ratio = -10 \times \log \left(\frac{y^2}{S^2} \right) \tag{13.3}
$$

where S is the standard deviation, y is the output average, y_i is the measured property, and n is the number of samples in each test trial.

13.1.5 Taguchi Method in Aesthetic Design

In spite of the fact that Taguchi method has been widely apply in functional design, there are a few researchers that have applied Taguchi method in aesthetic design. Initially, analysis in aesthetic design was performed using various methods in Kansei engineering. As market becoming more competitive, the application of Kansei engineering was regarded to be complicated and time consumed, due to the application of complex mathematical model which constructed over long period. Kansei engineering could not answer the challenge to provide rapid aesthetic design. This condition leaded researchers to a method that not only can provide product with a short development cycle time, but also able to comply with aesthetic diverse market. Taguchi method was finally selected to fill these gaps.

Amongst the works of a few researchers who applied Taguchi method in aesthetic design is the application of Taguchi method to improve the feeling quality [3, 4]. They work has successfully confirmed that Taguchi method can be applied in aesthetic design as effectively as in functional design. Our research is another work in the application of the robustness of Taguchi method in aesthetic design. Taguchi method is coupled with other methods, such as: factor analysis and neural network, to create a basis of interactive aesthetic design system. In this work, the robust approach is used to evaluate and enhance the aesthetic design of PET bottle shapes.

13.1.6 Kano Model

Aesthetic is one of the many qualities in a product that can make the product sells. In order to make customer happy and willingly to buy a product, industries must concern on how to integrate product quality with customer's satisfaction. Dr. Noriaki Kano [5] integrated product quality with the degree of product performance and the degree of customer satisfaction. The Kano model is useful for to provide the level of customer sophistication. Figure 13.1 shows the illustration of Kano model, in which the level of customer satisfaction is plotted in vertical axis, while the level of product performance is plotted in horizontal axis. By using this illustration, different type of customer needs can lead to different response. In correlation to aesthetic design, it is important to apply Kano model to recognize product performance which can improve customer aesthetic satisfaction. In order to satisfy customer, product performance must meet the three types of customer needs, as described below:

Fig. 13.1 Quality types based on kano model

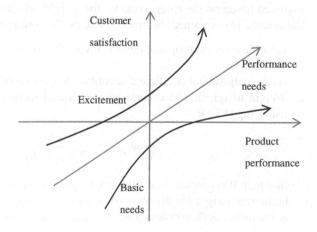

1. Basic needs: These needs should be achieved for customer satisfaction. If it not achieved, customers will be dissatisfied. Customers become dissatisfied if product performance is not achieved. However, product performance is not exceeding the neutral level of customer satisfaction.
2. Performance needs: product performance has a linear effect on customer satisfaction. The increase in product performance increases customer satisfaction.
3. Excitement needs: customer satisfaction increases in an exponential response with the increase of product performance.

[4] Added two more customer quality needs: the indifference and reversal qualities. The indifference need: the product performance is not affecting customer satisfaction, and the reversal quality: the increase in product performance makes customer more dissatisfied.

The categories of customer needs in Kano are identified using Kano questionnaire. Indication of customer satisfaction or dissatisfaction to product performance can be observed. The procedure for conducting Kano questionnaire consists of two stages. In the first stage, a scenario in which product parameter that gives performance is presented to respondent. On the second stage, the opposite condition of the first stage, respondent is presented with a scenario in which product parameter that give performance is excluded. For each stage, respondent must choose one of the options that express his/her level of satisfaction:

a. Satisfied
b. It should be that way
c. I am indifferent
d. I can live with it
e. Dissatisfied

By this means, criteria related to customer satisfaction can be categorized and product performances that lead to customer satisfaction can be given more priority. Criteria which fall in the excitement needs receive attention first, on the opposite criteria which fall in the basic need receive lower priorities. A weight adjustment is assigned based on the categorization; the weight adjustment is used to reprioritize the criteria. [4] explained the procedure to perform weight adjustment is as follow:

a. Identification of proper category for each criterion that related to customer satisfaction.
b. Assign adjustment coefficient according to Kano qualification classification.
c. Weight assignation to criteria with high and increasing customer satisfaction, using Eq. (13.1):

$$w_{i\_adj} = \frac{w_i K_i}{\sum_{i=1}^{n} w_i K_i} \tag{13.4}$$

In which $W_{i_{dj}}$ is the final adjusted weight for the ith performance criterion, W_i is the raw weight for the ith performance criterion, $i = 1, 2, 3\ldots$ n and K_i is the adjustment coefficient according to Kano quality classification.

d. The raw weight W_i is obtained from weighting method
   1. Convert S/N ratio value

$$z_{ij} = \frac{d_{ij}}{\sum_{j=1}^{m} d_{ij}} \quad (13.5)$$

where d_{ij} is S/N ratios for evaluation of the ith criteria in the jth experiment samples, $i = 1,2,\ldots,n$; $j = 1,2,\ldots,m$, then $D = (d_{ij})_{n*m}$ is the sample evaluation matrix.

2. Calculate the entropy of the ith criteria e_i

$$e_i = -c \sum_{j=1}^{m} z_{ij} \ln z_{ij} \quad (13.6)$$

In which $c = 1/\ln(m)$ and $e_i > 0$

3. Calculate the objective weight of the criteria

$$w_i = (1 - e_i) / (n - E) \quad (13.7)$$

where $E = \sum_{i=1}^{n} e_i$.

e. The adjustment coefficient K_i is value of 4, 2, 1, and 0, which are assigned to the excitement needs, performance needs, basic needs and indifference needs, respectively.

13.2 Methodology

The objective of the study is to find the relation between kansei words and PET bottle shape parameters. Kansei words are used to expresses customers' emotion and demand, and shape parameters are used to extract physical design which important for PET bottle shape images presentation. In this work, a Kansei engineering framework utilizing the robustness of the Taguchi method and the Kano model to enhance the aesthetic quality as shown in Fig. 13.2 is used.

PET bottle shape is selected to apply the method of robust design. PET bottle is selected because it represents the frequently purchased item by Japan citizens.

13.2.1 Collection of Kansei Words

In this step, as many as possible kansei words that represent the impressions evoked from PET bottle shape were collected. 26 PET bottles were presented to 18 university students with Japan national. Then, respondents gave opinion on the

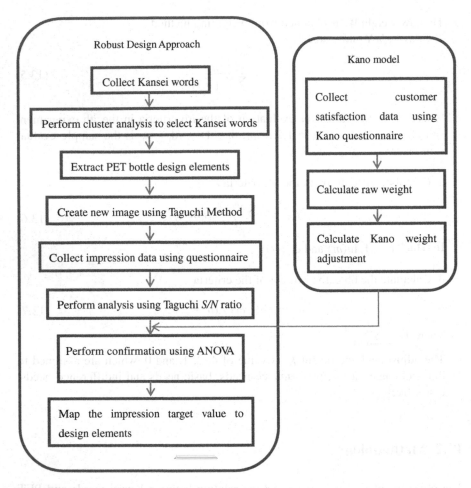

Fig. 13.2 Framework of the robust design approach

possible impression/emotion evoked from the bottle shapes. From this discussion 29 Kansei words were obtained. Table 13.1 shows the collected kansei words.

13.2.2 Cluster Analysis

In this step, important Kansei words were reduced from the large pool of kansei words collected in the previous step through discussion. Six bottles were selected randomly from the 26 bottles samples. The six bottles together with the 29 Kansei words were presented to respondents. Using all Kansei words, respondents were required to give score to each bottle shape. Every kansei words were scored from 1 to 5, using semantic differential scale [6, 7]. Score 1 meaning that the bottle shape

Table 13.1 Component matrix of the 29 kansei words

	Component			
	1	2	3	4
Unique	1.211	0.013	−0.048	0.113
Interesting	1.081	−0.133	−0.055	0.140
New	1.103	0.250	0.149	0.062
Standard	−1.165	−0.187	0.044	−0.259
Retro	−1.048	−0.248	−0.166	−0.060
Strange	1.104	−0.016	−0.152	0.153
Ordinary	−1.164	−0.224	−0.149	−0.315
Eye-catching	1.007	−0.014	−0.292	0.249
Simple	−0.991	−0.096	0.305	−0.267
Sweet	0.818	−0.162	0.374	−0.070
Cute	0.926	−0.158	0.498	−0.107
Healthy	−0.511	0.208	0.139	0.165
Fresh	0.712	0.195	0.411	0.143
Modern	0.697	0.454	0.072	0.208
Premium	0.610	0.483	0.193	0.346
Stylish	0.725	0.704	0.443	0.186
Feminine	0.692	0.537	0.664	0.093
Cool	−0.002	0.819	−0.027	−0.056
Clear	0.023	0.586	0.227	−0.102
Elegant	0.165	0.863	0.188	0.311
Adult	−0.269	0.746	−0.195	0.291
Childish	0.594	−0.904	0.365	−0.179
Slim	0.155	0.748	0.502	0.257
Yummy	0.252	0.365	0.133	0.007
Compact	−0.163	0.072	1.059	0.189
Dynamic	0.438	−0.214	−1.048	0.107
Masculine	−0.797	−0.213	−0.805	−0.128
Complex	0.540	0.256	0.088	0.748
Sophisticated	0.344	0.409	0.351	0.501

was lack of impression of presented Kansei words, and score 5 represent that the shape possess the impression presented Kansei words. The respondents consisted of 18 university students with Japan national, who's aged between 20 and 30.

The scored data was analyzed using exploratory factor analysis. Factor analysis was conducted using an extraction method based on the principle component analysis. Component with eigenvalues greater than 1 were extracted, resulting to 83.12 % cumulative contribution of three factors. Factor four was included on the extracted component with 10.1 % contribution of variance, making up a total of 96.22 % of cumulative contribution of variance. The extracted four factors were rotated using varimax rotation method. The results after varimax rotation are shown in Table 13.1.

Cluster analysis was used to devide the kansei words into four groups based on the rotated component matrix. Unweighted pair-group method using arithmethic averages (UPGMA) method was used for clustering method, Euclidean metric was selected in

UPGMA method. Agglomerative coefficient (AC) was used to measure the degree that samples are structured. AC shows the average of dissimilarities between clusters or samples when merged, in a standardized form [0, 1] [6]. Observation to an abrupt change of similarity/dissimilarity was performed on the AC. When samples or cluster merged, the similarity decreases (dissimilarity increase) monotonically. When different clusters were merged, it resulted in an abrupt change of dissimilarity (similarity). The number of cluster before gap was accounted for with a meaningful number. Figure 13.3 shows the graphic of AC and sample number. The dissimilarity when clusters are merged into one is shown in the rightmost graphic. Observing the graphic from right to left, we can find large gaps between first and second points (1 and 2 cluster), between second and third points (3 and 4 cluster), and between third and fourth points from the right. After that, a gentler gradient from the starting point can be observed. The fourth point from the right is the point which the dissimilarity starts to increase. The fourth point from right is termed as elbow. The number of cluster is determined by subtracting the number of samples with the sample in which *elbow* is occurred.

The number of samples is twenty nine, while the sample where elbow occurred is twenty five, resulting four as the number of cluster.

The member of the four clusters can be seen in Figs. 13.6, 13.7, 13.8, 13.9. In these figures, the features of clusters can be obtained by averaging the evaluation values of the members within the clusters. [6] selected Kansei words with the high average evaluation value from each cluster. He did not specify the cutting point of the average evaluation value to be selected as the features of cluster. In our study, the cutting point of the average evaluation value is set as three. Kansei words with average evaluation value equal or greater than three is selected to represent the cluster, as shown in Figs. 13.4, 13.5, 13.6 and 13.7. Eight Kansei words were obtained from the four clusters. Four pairs of Kansei words represent negative-positive relationship were constructed from the eight Kansei words (Childish–Adult, Masculine–Feminine, Ordinary–Stylish, Simple-Sophisticated).

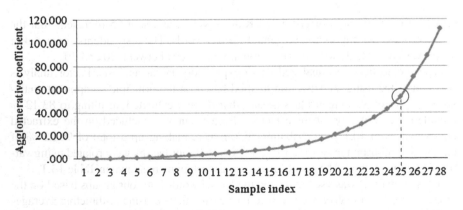

Fig. 13.3 Graphic of agglomerative coefficient

Fig. 13.4 Kansei words in cluster 1

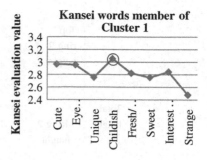

Fig. 13.5 Kansei words in cluster 2

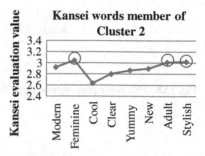

Fig. 13.6 Kansei words in cluster 3

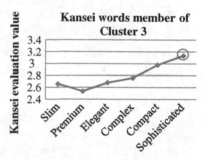

Fig. 13.7 Kansei words in cluster 4

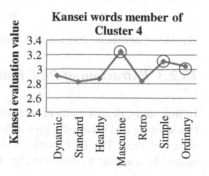

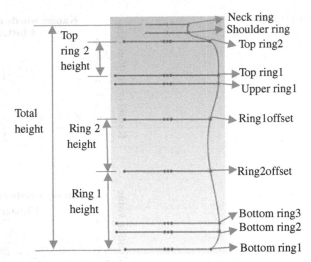

Fig. 13.8 PET bottle presentation image design elements

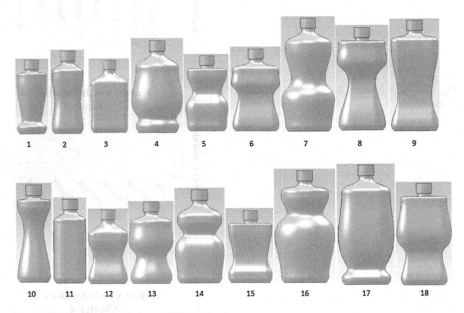

Fig. 13.9 Combinative design of PET bottle shapes

13.2.3 Extraction of PET Bottle Design Elements

In this study, design elements that effectively give different image design on PET bottle shape and eventually give different impressions to customers' product image were selected. The PET bottle design was created by several rings which connected by a spline, as shown in Fig. 13.10. The unique design of PET bottles is the result of the combination of various rings' diameter and their individual heights. Each design elements was evaluated, and important elements were selected for the study. Six design elements were extracted: Total height, Ring1 height, Ring2 height, Ring1 offset, Ring2 offset and Top ring2 height. Other design elements were set as constant. For each element, we set the parameter levels, as shown in Table 13.2. Parameterization for the design elements were determined by measuring the dimension of distributed PET bottles.

13.2.4 Creation of New PET Bottle Images Using Taguchi Method

In order to obtain optimized design, it is necessary to vary the design condition, so that, the relationship between design elements and the evoked impression can be observed. Taguchi orthogonal array is applied to generate combinative designs.

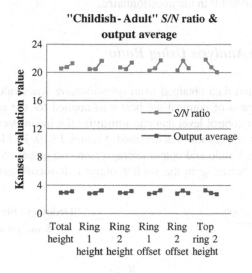

Fig. 13.10 Response graphs for *S/N* ratio of of "Childish–Adult"

Table 13.2 Design elements for PET bottle design

Design elements (DE)	Parameter level		
	1	2	3
Total height	150 mm	200 mm	250 mm
Ring 1height	0.1*total height	0.4*total height	0.7*total height
Ring 2 height	5 mm	10 mm	15 mm
Ring 1 offset	−5 mm	0	5 mm
Ring 2 offset	−5 mm	0	5 mm
Top ring 2 height	5 mm	15 mm	25 mm

The application of Taguchi orthogonal array required the designer to have controllable factor and uncontrollable factor. In our study, the controllable factor was taken from the design elements, while the uncontrollable factor was the inconsistency in customers' impression. Orthogonal array $L18$ was selected to generate combinative design, since it can accommodate six control factors with three levels. The six control factors were assigned from column 2 to 7. Figure 13.9 shows the generated combinative designs.

13.2.5 Data Collection Using Questionnaire

The questionnaire was designed based on the 18 combinative designs and the four pairs of kansei words. Each combinative design was assessed using the kansei words pairs. Respondents were requested to give impression scores from 1 to 5 using semantic differential scale [6]. As many as 21 university students with Japan national participated to fill in the questionnaire.

13.2.6 Perform Analysis Using Ratio

The kansei evaluation data obtained from questionnaire was evaluated using-Taguchi S/N ratio. S/N ratio of nominal the best was applied for the analysis. The optimal combination of control level that can minimize the inconsistency in customers' opinion of aesthetic impression was obtained. Figures 13.10, 13.11, 13.12 and 13.13 show the Taguchi S/N ratio and output average response graphs. S/N ratios and output averages were behaving in the similar manner. It seems that factors that can

Fig. 13.11 Response graphs for S/N ratio of "Ordinary–Stylish"

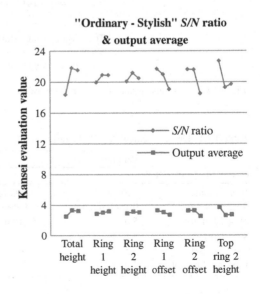

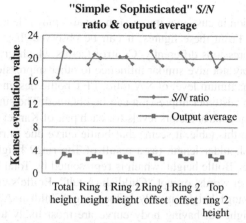

Fig. 13.12 Response graphs for S/N ratio of "Simple–Sophisticated"

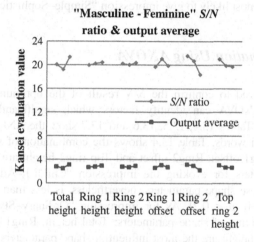

Fig. 13.13 Response graphs for S/N ratio of "Masculine–Feminine"

Table 13.3 Summary of optimum combination of control factor level

Impression	Optimal control level combination
Childish–Adult	Ring1 height (level 3), Ring1 offset (level 3), Ring2 offset (level 2) and Top ring2 height (level 1)
Ordinary–Stylish	Total height (level 2), Ring1 offset (level 1), Ring2 offset (level 1) and Top ring2 height (level 1)
Simple–Sophisticated	Total height (level 2), Ring1 offset (level 1), Ring2 offset (level 1) and Top ring2 height (level 1)
Masculine–Feminine	Total height (level 3), Ring1 offset (level 2), Ring2 offset (level 1) and Top ring2 height (level 1)

minimize the variation in customer impression are also play role to obtain the desired target mean value. From these figures, it can be observed that control factors are influencing the impressions differently. Control factors which give strong influence to an impression may not give similar influence to other impressions. By setting the control factors at maximum level of *S/N* ratio, PET bottle design that can minimize the inconsistency in customer impression can be obtained. The summary of the optimum combination of control factors levels for each pair of Kansei words is presented in Table 13.3. From this table, it seems that bottle curve due to rings (Ring1 offset, Ring2 offset) and shoulder length (as a result of Top ring2 height) are influencing for all Kansei words. Bottle height, which is represented by Total height, is influencing almost all Kansei words, except "Childish–Adult". Bottle's curve height, which is represented by Ring1 height, is only influencing "Childish–Adult". In summary, bottle with low height but having body curve are most likely to give impressions "Ordinary–Stylish", and "Masculine–Feminine", while bottle with medium height and body curve is most likely to give impression "Simple–Sophisticated".

13.2.7 Confirmation Using ANOVA

ANOVA was utilized to confirm the *S/N* result of the optimum combination of control factors. ANOVA can identify factors which significantly influence customer impression. Tables 13.4, 13.5, 13.6 and 13.7 show the ANOVA result for the four pair of kansei words. Table 13.4 shows the combination of shape parameters: Ring1 height, Ring1 offset, Ring2 offset and Top ring2 height are the most influential shape parameters for evoking the impression "Childish–Adult". The significance effect of these shape parameters, nevertheless, is less than 90 % confidence level (approximately $p < 0.2$). As for impression "Ordinary–Stylish", Table 13.5 shows the combination of shape parameters: Total height, Ring1 offset, Ring2 offset and Top ring2 height are the most influential shape parameters with significance effects: $F(2,4) = 10.08, p < 0.05; F(2,4) = 4.11, p < 0.2; F(2,4) = 9.07, p < 0.05$, and $F(2,4) = 17.15, p < 0.05$, respectively. It is observed that only Ring1 offset is having

Table 13.4 ANOVA table for "Childish–Adult"

Source	SS	dof	MS	F	%Cont.
Total height(e)	0.14	2.00	0.07		
Ring 1 height	0.70	2.00	0.35	3.08	13.03
Ring 2 height(e)	0.32	2.00	0.16		
Ring 1 offset	0.79	2.00	0.39	3.47	15.46
Ring 2 offset	0.71	2.00	0.35	3.12	13.26
Top ring 2 height	0.98	2.00	0.49	4.31	20.70
Error pool	0.46	4.00	0.11		37.56
TOTAL	3.64	12.00			100.00

Table 13.5 ANOVA table for "Ordinary–Stylish"

Source	SS	dof	MS	F	%Cont.
Total height	2.42	2.00	1.21	10.08	19.55
Ring 1 height(e)	0.31	2.00	0.16		
Ring 2 height(e)	0.17	2.00	0.09		
Ring 1 offset	0.99	2.00	0.49	4.11	6.70
Ring 2 offset	2.18	2.00	1.09	9.07	17.38
Top ring 2 height	4.12	2.00	2.06	17.15	34.78
Error pool	0.48	4.00	0.12		21.59
TOTAL	11.16	12.00			100.00

Table 13.6 ANOVA table for "Simple–Sophisticated"

Source	SS	dof	MS	F	%Cont.
Total height	6.86	2.00	21.99	72.38	57.15
Ring 1 height(e)	0.81	2.00	4.58		
Ring 2 height(e)	0.43	2.00	2.99		
Ring 1 offset	0.99	2.00	3.57	11.75	3.47
Ring 2 offset	0.78	2.00	3.11	10.24	1.60
Top ring 2 height	1.07	2.00	4.97	16.34	4.24
Error pool	1.22	4.00	0.30		33.54
TOTAL	10.94	12.00			100.00

Table 13.7 ANOVA table for "Masculine–Feminine"

Source	SS	dof	MS	F	%Cont.
Total height	1.24	2.00	0.62	6.42	17.71
Ring 1 height(e)	0.21	2.00	0.11		
Ring 2 height(e)	0.18	2.00	0.09		
Ring 1 offset	0.73	2.00	0.36	3.76	9.02
Ring 2 offset	2.94	2.00	1.47	15.23	46.48
Top ring 2 height	0.62	2.00	0.31	3.21	7.21
Error pool	0.39	4.00	0.10		19.59
TOTAL	5.91	12.00			100.00

significance level less than 90 % confidence level. While for impression "Simple–Sophisticated", Table 13.6 shows the combination of shape parameters: Total height, Ring1 offset, Ring2 offset and Top ring2 height are the most influential shape parameters with significance effects: $F(2,4) = 72.38, p < 0.001$; $F(2,4) = 11.75, p < 0.05$;

$F(2,4) = 10.24$, $p < 0.05$; and $F(2,4) = 16.34$, $p < 0.05$, respectively. And finally, for adjective word pair "Masculine–Feminine", Table 13.7 shows the combination of shape parameters: Total height, Ring1 offset, Ring2 offset and Top ring2 height are the most influential shape parameters with significance effects: $F(2,4) = 6.42$, $p < 0.1$; $F(2,4) = 3.76$, $p < 0.2$; $F(2,4) = 15.23$, $p < 0.05$; and $F(2,4) = 3.21$, $p < 0.2$, respectively. The confidence level of Ring1 offset and Top ring2 height is less than 90 %.

It seems shape parameters give different influence to evoke impressions. It is observed that the diameter in Ring1 offset and Ring2 offset, and the height between shoulder and body play an important role in giving impressions of all adjective words to customers. Bottle's Total height seems to be affecting the impressions of "Ordinary–Stylish", "Simple–Sophisticated" and "Masculine–Feminine", while Ring1 height seems to be evoking the impression of "Childish–Adult". Ring2 height is not giving influence for evoking all impressions.

In spite that the combination of shape parameters for "Ordinary–Stylish", "Simple–Sophisticated" and "Masculine–Feminine" are same, the level combination of the shape parameters could not be determined using ANOVA. From these tables, ANOVA indicates that the shape parameters have a different average results from level to level.

Comparison to the combination of control factor that affecting Kansei words from Taguchi *S/N* ratio and ANOVA was performed. The comparison shows that ANOVAs present the similar result to *S/N* ratio result, despite that ANOVA cannot identify the control factors' level. This implies that ANOVA confirm the results from *S/N* ratio.

13.2.8 Mapping of Kansei Weight to Design Elements

In the previous section, it can be observed that Taguchi method was able to recognize the optimal control level at fix parameter level number. In this section, a basis of interactive aesthetic system design was created by utilizing factor analysis and neural network (NN). Figure 13.14 shows the flow in robust design method when applied in interactive aesthetic design. The aesthetic target values represented by Kansei word were map to design parameters. Range of the aesthetic target value was made from 1 to 5. Factor analysis and neural network were applied for the purpose of mapping.

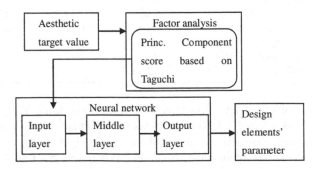

Fig. 13.14 Mapping aesthetic value to design parameters

Table 13.8 Principal component analysis

Component	Initial eigenvalues			Extraction sums of squared loadings		
	Total	% of variance	Cumulative %	Total	% of variance	Cumulative %
1	2.141	53.517	53.517	2.141	53.517	53.517
2	1.028	25.708	79.225	1.028	25.708	79.225
3	0.658	16.439	95.665	0.658	16.439	95.665
4	0.173	4.335	100.000			

Table 13.9 Rotated component loading matrix

	Component		
	1	2	3
Simple–Sophisticated	0.973	0.026	−0.041
Ordinary–Stylish	0.779	0.471	0.272
Masculine–Feminine	0.143	0.970	0.144
Childish–Adult	0.057	0.150	0.984

Factor analysis was used to reduce the dimensionality of the data and also to eliminate data with less variance contribution. In this way, data that really reflecting the customers' Kansei can be obtained. First, design parameters' level with high S/N ratio were selected, after that, the output average of each design parameter were summed. If the summation is less/more than the target value, then less important design parameters' level were changed to the second highest S/N ratio. Then, the summation of the design parameters was compare to the aesthetic target value. This process was repeated until the target value is achieved. The design parameters subject to change the level are the design parameters which insensitive to the inconsistency of customer impression.

S/N ratio and output average data were used for factor analysis. Components with eigenvalue greater than 1 were extracted, resulting on 79.23 % of cumulative variance contribution. Factor 3 has eigenvalue less than 1, nevertheless, it contributed to 16.44 % of variance, and therefore was included in the extracted factors. The cumulative contribution of the three factors was 95.67 % as shown in Table 13.8. The result after varimax rotation is shown in Table 13.9. Principal component score can be obtained from multiplication result of kansei weight and component loading matrix.

Multilayer feed forward back propagation neural network was used to map principal component score to design elements parameter. The input for the neural network was the principal component score, and the output of the network was the design elements parameter. The architecture of the neural network consisted of three nodes in the input layer, ten nodes in the hidden layer, and six nodes in the output layer. Table 13.10 shows the mapping result of the NN when Kansei points in Table 13.11 is inputted. From this table, it can be observed that the accuracy for neural network to map the aesthetic value to design parameters is 50–75 %. The neural network was trained with a small number of data, since only 21 data were obtained during the questionnaire. Training the network with more data might improve the accuracy of the network.

Table 13.10 NN result

Impression	Optimal control level dimension
Childish–Adult	**Ring1 height** (0.6998 mm), **Ring1 offset** (5 mm), Ring2 offset (5 mm) and Top ring2 height (24.95 mm).[50 % accuracy]
Ordinary–Stylish	Total height (1.0 mm), **Ring1 offset** (−4.52 mm), **Ring2 offset** (−4.99 mm) and **Top ring2 height** (5.04 mm). [75 % accuracy]
Simple–Sophisticated	Total height (1.0 mm), **Ring1 offset** (−5.42 mm), **Ring2 offset** (-4.99 mm) and **Top ring2 height** (5.04 mm). [75 % accuracy]
Masculine–Feminine	Total height (1.0 mm), Ring1 offset (−4.52 mm), **Ring2 offset** (−4.99 mm) and **Top ring2 height** (5.04 mm).[50 % accuracy]

Table 13.11 Kansei words' weight

Kansei columns	Target impression weight			
	Adult	Stylish	Sophisticated	Feminine
Childish–Adult	5	1	1	1
Ordinary–Stylish	1	5	1	1
Simple–Sophisticated	1	1	5	1
Masculine–Feminine	1	1	1	5

13.3 Kano Model

Kano model is used to enhance the quality feeling of the aesthetic design. Customer aesthetic satisfaction can be improved using Kano model. The quality feeling is enhanced by multiplying design elements parameter to Kano weight.

The Kano weight was obtained from Kano questionnaire. Kano questionnaire was created and presented to 11 respondents of university students with Japan national. In Kano questionnaire, the aesthetic criteria were classified based on customers' aesthetic needs. Table 13.12 shows the result of the Kano questionnaire which provided information for classifying criteria. The criteria of "Ordinary–Stylish" and "Masculine–Feminine" can be considered excitement needs; "Childish–Adult" and "Simple–Sophisticated" can be considered performance needs. Efforts should be directed toward the at excitement and performance needs. The Kano classification corresponding to each criterion was the integrated to adjust the criteria weight by multiplying the adjustment coefficient (K) with each Kano category.

S/N ratio of the four aesthetic criteria was used for calculating raw weight criteria. Equations (13.5–13.7) were used to calculate the raw criteria weight. Adjustment of each criteria weight according to its Kano category was performed by integrating the Kano model to the raw weight criteria. Equation (13.4) was used to calculate the final adjusted weight, which explaining about the prioritization related to customer satisfaction. Table 13.13 shows the result of the adjusted weight incorporating with Kano model. These results imply that excitement needs,

Table 13.12 Kano questionnaire result

Aesthetic criteria	E	P	B	I	R	Kano category
Childish–Adult	3	7	0	0	1	P
Ordinary–Stylish	2	2	0	2	5	E
Simple–Sophisticated	1	6	0	2	2	P
Masculine–Feminine	6	4	0	0	1	E

E excitement, *P* performance, *B* basic, *I* indifference, *R* reversal

Table 13.13 Kano weight for aesthetic impression

Impression	Childish–Adult	Ordinary–Stylish	Simple–Sophisticated	Masculine–Feminine
Kano weight	P	E	P	E
K	2	4	2	4
Raw weight	0.339961	0.339786	0.339743	0.339903
Adjusted weight	0.166722	0.333274	0.166615	0.333388

such as "Ordinary–Stylish" and "Masculine–Feminine", must be prioritized. "Childish–adult" and "Simple–Sophisticated" were categorized as performance needs, and therefore received the second priority.

13.4 Development of Aesthetic Shape Generation System

Aesthetic generation system was developed as an intelligent tool inside Solidworks through the use of its application programming interface (API). The APIs are implemented by writing function calls in the program, which provide linkage to the required subroutine for execution. In Solidworks, the API is written in Visual Basic language with Microsoft Visual Basic for Application (VBA). Microsoft VBA is a toolset based on Microsoft Visual Basic 6.0 and is embedded inside Solidworks software. It enables automation process by calling Solidworks functions from the code writing on VB application. The aesthetic system was built based on the robust design approach and Kano model.

Figure 13.15 shows the interactive aesthetic evaluation interface.

The system can help designer to make an aesthetic PET bottle design designer aesthetic intention. The integration of Kano model to the robust design enables the system to enhance the aesthetic criteria. First, designer input the target impression for each pair of Kansei words. Then, by clicking the button "Calculate parameters", system calculates the aesthetic weight to the appropriate value of design parameters. System also calculates the Kano adjustment weight for aesthetic design enhancement. Finally, by clicking the button "Generate model", system generates the aesthetic PET bottle model according to designer target impressions.

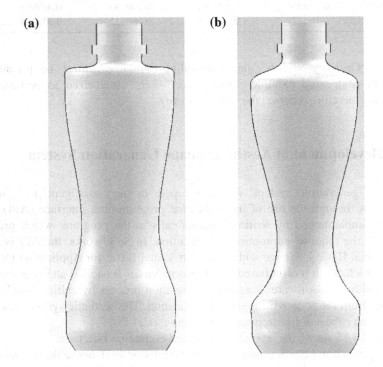

Fig. 13.15 PET bottle interactive system design

Fig. 13.16 Aesthetic model enhancement using Kano model. **a** Before kano model analysis. **b** After kano model analysis

Figure 13.16 shows the PET bottle model before generated by the system (a) before integrating Kano model and (b) after integrating Kano model to the robust design approach. The bottle shape was generated when designer was inputting the target value of "5" for all Kansei word pairs. Visual comparison between PET bottle design before and after the application of Kano model shows that the design after the application of Kano received enhancement on the design parameter of Ring1 offset and Ring2 offset, which cause the model look more appealing in terms of the four pairs of Kansei words. From these figures, it can be observed that Kano model is enhancing the aesthetic impression.

13.5 Conclusion

This study attempts to study how PET bottle shapes affecting customers' impressions. This study analyzes the relationships between customers' emotion and PET bottle design elements using robust design approach. The study utilizes robust design approach and Kano model to optimize the quality of aesthetic satisfaction. Robust design method has been applied to obtain the optimum combinations of design elements that closely match to customers' desired impressions. The application of robust design helps designer to identify design elements that important for influencing impressions. By obtaining optimal combinations of design elements levels, design which stable/independent to the inconsistency of customers' impression can be achieved. Result of the study shows that customers Kansei responses consisted of four emotional factors. It was also shown that bottle's height, bottle's curve and curve's height play important role in giving different impressions to customers. The result can be used to evaluate and enhance future PET bottle shape presentation image designs. The Kano model enhanced the aesthetic design by giving emphasis on the aesthetic criteria that related to customer satisfaction.

An interactive system design has been built based on the robust design and Kano model approaches. The system was able to assist even non designer to create a personal PET bottle shape that closely satisfy his/her desired impressions. The robust design approach can also be applied to other product design.

References

1. Shuichi F (2011) Emotion: a gateway to wisdom engineering. In: Shuichi F (ed) Emotional engineering. Springer, London
2. Phillip JR (1988) Taguchi technique for quality engineering. McGraw-Hill, New York
3. Hsin HL, Yu MC, Hua CC (2005) A robust design approach for enhanching the feeling quality of a product: a car profile case study. Int J Ind Ergon 35:445–460. doi:10.1016/j.ergon.2004.10.008
4. Chun CC, Ming CC (2008) Integrating the kano model into a robust design approach to enhance customer satisfaction with product design. Int J Prod Ergon 114:667–681. doi:10.1016/j.ijpe.2008.02.015

5. William YF, Clyde MC (1995) Engineering methods for robust product design using taguchi methods in technology and product development. Addison Wesley, New York
6. Mitsuo N (2011) Kansei/affective engineering. CRC Press Taylor and Francis Group, Roca Baton
7. Mitsuo N, Anitawati ML (2011) Innovations of kansei engineering. CRC Press Taylor and Francis Group, Roca Baton

Chapter 14
Multisensory User Experience Design of Consumer Products

Monica Bordegoni, Umberto Cugini and Francesco Ferrise

Abstract In recent years, the production of consumer goods has exceeded that of industrial products. This has led to changes in the areas of design and production. The target users of industrial products (in the Business to Business—B2B market) are industries that decide to purchase a product on the basis of its technical features, functions and performance. Differently, the target users of consumer products (in the Business to Consumer—B2C market) are the consumers who choose a product driven by other aspects, besides features and functions, such as the perceived value, the expected benefits, the emotions elicited, as well as features and functions. This has brought a paradigm shift in the design process. And in fact, the design of consumer products is increasingly focusing on the so-called *user experience*. The designer must not design only the product, but also the user experience in relation to its use. The resulting product should have a high perceived value and generate positive emotions in the consumer. These factors make a successful product on the market. Therefore, the role of the designer is designing the products and the perceptual aspects of their use, that is, designing the user experience and deriving from that the products' specifications. Consequently to that, the design process has changed in the last years. In fact, the user is now at the center of the design process, named user-centered design. Being the new focus the target users, the evaluation of their interaction with the new designed products is expected to be rigorous and systematic. An efficient approach has proved to be one in which the validation is made up by involving users in the early stages of the product design. Since typically at the level of the concept the product, or a prototype that is comparable with it at the perceptual level, is not available, it is not possible to make a thorough validation of its use with users. However, new methodologies of Virtual Prototyping allow us to simulate multisensory user interaction with product concepts early in the design process. This chapter introduces the use of interactive Virtual Prototyping (iVP) methodology for the design of the user experience with the concept of a new product. Interactive components of new products and their behavior are simulated through functional models, and users can experience them through multisensory Virtual Reality (VR) technologies.

M. Bordegoni (✉) · U. Cugini · F. Ferrise
Dipartimento di Meccanica, Politecnico di Milano, Via La Masa 1, 20156 Milan, Italy
e-mail: monica.bordegoni@polimi.it

14.1 Introduction

Product innovation refers to the design and implementation of newly designed products, which are proving successful in the market for their novel aspects, which may concern aesthetics, functionality and usability. Product innovation is based on a first generative phase in which many solutions are proposed in response to an initial market question [1, 2]. A creative approach of the designers and engineers at this stage is crucial in order to identify genuinely new solutions. Obviously, the creative phase is also very exploratory, and the set of solutions identified must then be carefully analyzed in order to discard most of those and preserve the one to implement.

The traditional approach of design is based on the proposition of *engineering* (*technical and quantitative*) *questions* that focus the research of the design solutions on issues relating to the efficiency and effectiveness of the product, in relation to the task it must perform.

A product is designed and produced for its consumers, and this is especially true for consumer products, that is, those goods intended for direct use or consumption by the average consumer. Consumers buy a specific product for several reasons, for example, because of its price or brand. Or for its features, for its specific functionalities, but also for the perceived benefits the product delivers. These aspects are related to the perceived value of a product, which can be defined as the mental estimation a consumer makes of it [3]. When considering consumer products, the consumption experience is the true value of a product for consumers. In this view, the consumers, and all issues related to them like feelings, preferences, emotions, become central in the design of a new product.

Actually, the traditional design approach based on engineering questions is independent of considerations related to the consumer. The questions that the designer must answer concern engineering aspects, such as Does the solution work? Are consumption and cost optimized? Is the product sturdy, durable, etc.? Such questions certainly do not consider the preferences of the user of the product. In fact, questions that can be defined as *emotional questions* are not typically raised, such as Does the user like the product for what concerns its aesthetics, usability, comfort? Which are the user preferences? Does he prefer solution *A* or solution *B* or rather *C*?

As it is demonstrated that emotional aspects contribute greatly to the success of a product and to the affection of the users for that product or brand at long term, today's design cannot ignore the study of the emotional aspects, since the early stages of the product concept [4]. The bond of an individual to a product is significantly driven by the emotional-subjective values. Actually, emotion is one of the strongest differentiators in user experience principally because it triggers unconscious attitude and mood in relation to a product [5].

Nowadays, users have clear expectations about a product. For example, they assume that a product works properly and also that they have the possibility of choosing among many comparable alternatives at the moment of purchase. What is becoming evident is the importance that the user (a potential customer before becoming a user) really likes the product, from various perspectives, and chooses it among a set of alternative products offered by competitors. So, it is important to ensure that the product captures the preferences of the user through several features, such as form, finishing, material, meaning and others.

Therefore, generating positive emotions in consumers becomes an essential design issue. In the view of designing a successful product, the designer should now favor the users' preferences, rather than the product performances. And this is why emotion becomes one of the key requirements of design. As the emotional responses should be predicted already during the conceptual phase, the drivers of the emotions can be intended as tuner for design: an elicited negative emotion can be used to change the product features as long as a positive emotion is yielded.

Engineering questions usually generate answers and solutions that are well analyzed, observable, measurable and quantifiable. Instead, how to understand and consider the implicit and explicit desires and feelings of the individual, how to measure the emotions and how to trigger them when interacting with the product are actually open issues. An additional issue is also about how to test the design solutions, considering the emotional aspects and not just the functional ones. Or indeed, the question to answer is about how to organize the design requirements around the emotions that embody users' expectations and preferences.

Users learn about products through a direct experience, which happens through an actual interaction with the product. People interact with everyday objects through sensory dimensions, which include vision, touch, hearing, smell and taste. This experience is also important when purchasing a product. Actually, it is also on the basis of the kind of experience one makes that one chooses and buys a product rather than another one.

When considering the interaction between a user and a consumer product, the sense of touch is a fundamental part of our experience. In fact, everyday, we use products and we also use our sense of touch as an integral part of our experience when using these products. It has been demonstrated that the design requirements should also address haptic interaction with products, including the user emotional engagement [6]. Haptic interaction design (HID) is then a topic of major importance deserving study and experimenting [7], as well as sonic interaction design, since touch also works in tight partnership with vision and hearing in many ways [8]. So, it is necessary to consider the user interaction with the product as a *multisensory* experience.

This chapter first describes the evolution of the design context, where in the last decades we have assisted to an evolution of machines, systems, products and services, and accordingly the changes in the design methodology and paradigms, in the role of designers, and in enabling tools as well. Then, the chapter reasons about the relationship between consumers and products, including perceived value of a product and user interaction with a product. The two subsequent sections present issues about user experience design (UXD) and how iVP can be used for the simulation of the user experience, including two case studies as examples. Finally, the last section presents the conclusions.

14.2 The Evolution of the Design Context

In order to fully understand the present design practice and its evolution trend, it is useful analyzing the evolution of design, also in relation to the kinds of artifacts it addressed. This section presents the evolution of the design context where we analyze the objective and output of the design, the design process, the role of designers and the enabling design methods and tools. These issues are tightly correlated; for in the last decades, the design process has evolved, in accordance with the evolution of the typology and complexity of the artifacts, and new tools have been introduced, and also designers have taken on new roles.

14.2.1 Machines, Systems and Products

Traditionally, the designers' activity was about creating and designing machines. Machines were intended as devices consisting of fixed and moving parts that modify mechanical energy and transmit it into a more useful form [9]. Therefore, machines were built with the aim of satisfying functions, generally oriented to help humans in doing things usually amplifying their craftsmanship and skill, by decreasing risk, fatigue, etc. In that time, users' expectation was that the machine operates properly and does not break.

Besides machines, systems have been subjects of study for design teams. Systems are more complex than machines, as they consist of groups of interacting, interrelated elements forming a whole complex. Systems are characterized by being multitechnological and with embedded controls and usually can be made up of mechanical electronic, pneumatic and oleodynamic components. Systems consist of assembly of subsystems, based on varied technologies for operating as a whole. Also for complex systems, the expectations of users are essentially related to technical functions, with an emphasis on reliability, fault tolerance, adaptability, flexibility, etc.

While technical systems can be part of a whole without any relation with humans, and only interact with other technical systems, products are strictly related to their users. A product can be defined as anything that can be offered to customers, which might satisfy a want or a need. The increasing spread of products with respect to systems has been accompanied by the growing importance of the users in relation to the products. Their satisfaction about the product is of primary importance for its success. Besides, designing a product also means designing the market addressed by the product and the accompanying services [10, 11].

In summary, the subject of design includes machines and systems, which have little to do with users, and products, which conversely have a strong relation with their users.

14.2.2 Functions and User Experience

In case of machines or systems, the users' main interest is focused on issues concerning their functions and technical features. Actually, it is important that the system that works well, does not break, is efficient and performing. Accordingly, the design is primarily aiming at ensuring these characteristics.

But in recent years, the focus has shifted. In fact, innovation in design is also crucial when delivering systems and industrial products more generally. Innovation has to do with the improvement of existing products or with the development of new products, which have new features, new aesthetics, new materials, etc. The technological innovation is accompanied by an additional value, namely the *user experience*, which is related to the effects that the product's use have on users.

These effects are particularly important in relation to the first contact that a customer has with a product during the purchasing decision process. Most of the times, the customers choose and buy what they like most. The motivation is often not clearly expressed or expressible and is somehow related to the emotional response, which induces the user to prefer a product to another.

In this changing and evolving context, also the design process has evolved accordingly, as described in the following section.

14.2.3 The Evolution of the Design Process

The design process has followed the evolution of the typology of designed artifacts. At least four types of design processes and approaches can be identified, which are related to the kind of users' needs (explicit or unspoken) and to the kind of products (incremental or new) (Fig. 14.1). The first are the design questions, and the second are the answers to the design questions, that is, the design solutions.

1. *Explicit needs versus incremental products.* The design of a product typically answers to a question derived by the analysis of the users' needs. The design can be very traditional and proposing an incremental development of a product built on previous products, engineering solutions and components. This kind of design process traditionally includes several phases aiming at obtaining a definition of a product that is detailed enough and also optimized for the following manufacturing phase [12]. These phases include conceptual phase and detailed design phase, evaluation phase and redesign and optimization phase. This approach is very typical and somehow conservative and is adopted by several companies developing products based on incremental solutions that answer to present users' needs.

2. *Explicit needs versus radically new products.* The design activity can bring to the definition of new products, which answer to questions explicitly posed by a group of target users. In order to make the proposed product a successful one, the design must be a response to an actual user demand. This approach is typical of

Fig. 14.1 Kinds of products versus users' needs

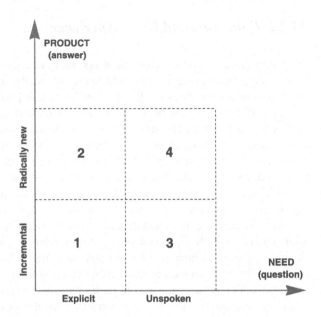

technology-push paradigm, where the design solutions developed can be intended as answers for the current traditional market or can even open new markets [13]. For example, the PC laptops were new products based on an incremental design that at that time satisfied the present market; instead, video game consoles were new products, creating the very new market of home video games back in the early 1980s.

3. *Unspoken needs versus incremental products.* The design can also be based on latent and unspoken needs that are somehow described, typically by marketing experts. Traditionally, users' needs and requirements are collected and identified through surveys that are carried out by companies' marketing departments. Many studies have shown that user's needs surveys based on customers' statements do not express real customer needs [14, 15]. Currently, in order to overcome these problems, some other methods are used. For example, ethnographic methods are used to study the behaviors of a small number of testers when using products. However, some issues still remain, despite attempts to improve survey methods for users' needs collection. These issues relate to the difficulty in collecting data about "unspoken" customer needs, which could represent the most important input for defining the requirements of new and innovative products. Once the unspoken needs are identified, the designers proceed in the search for possible solutions to the questions posed by the requirements. Since the solutions can be many and varied, the designs require performing careful tests, followed by optimization, in order to converge to the final and optimum solution. Products answering to unspoken needs can be suitable for the present market. An example is that of the electric city cars. Besides, there are products that open new markets. An example is the Roomba vacuum cleaning robot,

which is not a totally new product considering its functionality, but has certainly opened a new market in the field of household appliances [16].

4. *Unspoken needs versus new products.* Finally, a new design can be developed for satisfying unspoken needs. This can lead to the design of completely new products. Given the uncertainty of users' acceptance, the product has to be developed by holding the users in high regard and often involving them in the validation phases of the design choices. Actually, the level of success is less predictable than in the previous cases, and therefore, a more careful analysis and testing is required. The Segway is an example of an innovative product designed for satisfying unspoken users' needs, which has been created for a traditional market that is the one of personal mobility [17]. Instead, recent examples of innovative products created for a completely new market are personal 3D printers, which allow us to easily and directly create physical objects from 3D models [18].

14.2.4 People Responsible for the Design Results

Within the evolving context presented so far, also the role of the people responsible for the design process has changed in years (Fig. 14.2). Various skills are required to manage the various phases in the design process.

The roles of the various expertise involved in the design process can be listed as those who

- define the requirements
- propose the solutions
- decide and organize the validation

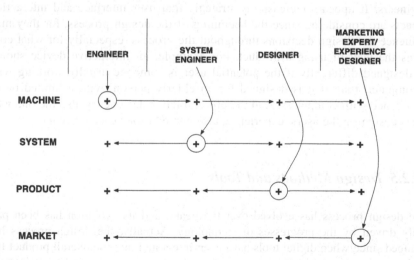

Fig. 14.2 Roles of people involved in the design with respect to the evolution of the design subjects (machine, system, product, market)

- validate the product
- choose among the various alternatives during the process and are in charge and responsible for the design as a whole

The design of machines was mainly carried out by *mechanical engineers* who approached the design problems from a functional point of view and often starting from known solutions. With the increasing complexity of machines, the role of designers has become multidisciplinary. Actually, design teams have replaced designers, and the *system engineer* is the one who designs the architecture of complex systems and who coordinates the design and assembly of subsystems.

The requirements of modern products are not primarily and uniquely related to engineering and technical questions, and the design process does not follow the traditional sequence of mechanical product design, as well as the role and kinds of practitioners involved have changed. In the view of designing products that are successful with users, it is important that designers always keep in mind who the end user of a product will be. *Industrial designers* have key roles and are sort of product architects, the ones who collaborate with several experts and have a whole vision of the product. They are called to design innovative solutions to real market problems and to define the overall product architecture and are expected to work closely with product and marketing managers, user interaction designers and software engineers to develop new products and to improve existing ones. This new figure of designer is knowledgeable about design (shape, aesthetics, ergonomics) and engineering (function, materials and production).

It is quite typical of the consumer market that *marketing people* give input to the design process of new products, by defining the initial requirements. Then, those are transformed into other kinds of requirements by the industrial designers, and by the system engineers, and eventually are handled by the mechanical engineers. It appears increasingly strategic that user interface and interaction aspects are considered since the beginning of the design process, for they may influence the design decisions throughout the process, especially for what concerns the way of using a product. For example, an interactive device should be designed differently if the potential user is someone usually working with a computer, than if it is designed for an elderly person, with a limited or no experience with computers. And therefore, today's designer is also the one who starts designing the users' experience, also named *experience designer*.

14.2.5 Design Methods and Tools

The design process has evolved over the years, and its evolution has been partially driven by the progresses of technology. Actually, the design process has changed since when digital tools have been integrated into the overall product life cycle management. At the beginning of the introduction of digital tools into the design process, design tools were used for replicating the production of design deliverables, that is, drawings. Today, digital tools are more intended to integrate and collaborate with the designers so as to expand their joint capabilities used to design and also test the results of the design activity.

Figure 14.3 shows a diagram reporting design methods versus tools. Design methods are classified as function-based and use-based, and design tools as geometry-centered and interaction-centered. Hereafter, we describe in details the evolution of digital methods and tools for product design and testing, including Digital Mock-up, Virtual Prototyping and iVP.

14.2.5.1 Digital Mock-Up: Function-Based: Geometry-Centered

The activities concerning the design of products are supported by various tools, which are mainly based on a geometric representation of the product and aiming at solving engineering and functional problems. Computer-aided design (CAD) tools support the geometrical description of products, including details as dimensions and tolerances.

Digital Mock-Ups (DMU) have been introduced to simulate and test the behavior and the performance of the components of products designed by using CAD tools. DMU are typically geometry-based, where the geometry of the system, subsystems and components is precisely defined [19]. The features describing the product, which are necessary for assessing its behavior and performances, are attached to the geometry. The DMU is therefore a static representation of the product or system, which can be used for simulating the movement of the various parts with the aim of detecting clashes, collisions and interferences, and simulating assembly and disassembly activities.

Fig. 14.3 Design methods versus tools

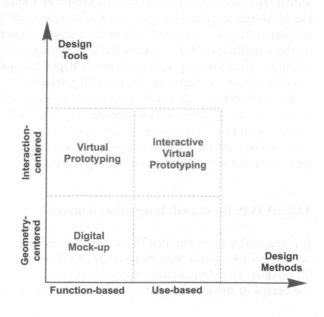

Various functional models can be derived from the DMU, which can be used for testing specific aspects of a product, for example, structural models, dynamic models. Much of the embodiment and detailed design phases of mechanical design consist of ensuring that the components in the design will not fail mechanically, either through excessive stress or through excessive deformation. Initially, this is done by using simple mathematical models. Then, more elaborate analysis methods are used, such as finite element analysis or computational fluid dynamic analysis.

The results of each test and analysis are used to solve a specific design problem, which does not necessarily mean optimizing the overall design of the product. In fact, the optimization of one aspect made on the basis of the analysis results may impact on other aspects. Therefore, there is a great need for more comprehensive testing methods.

14.2.5.2 Virtual Prototyping: Function-Based: Interaction-Centered

Virtual Prototyping is instead a common practice used to test the design solutions mainly with the aim of evaluating product function performance [20]. Virtual Prototyping consists of a combination of single- or multidomain functional models, which fully represent the physical behavior of a product, and VR technologies, which are used to give realism to the product representation. The geometry of the product can be simplified and is not anyhow the core of the prototyping; on the contrary the functional models are. The functional models can be modified during the testing phase, and the results are evaluated globally, thus providing a comprehensive view of the product.

In addition, VR technologies are used to realistically represent and render the virtual prototypes to users [21]. Most of Virtual Prototyping has been based on the use of visualization technologies, such as stereoscopic displays. In a sense, they are sort of "visual prototypes," where the realism of the representation is focused on the visualization of the product features. In order to overcome this limitation, recently Virtual Prototyping has integrated haptic technologies for simulating the physical contact and haptic interaction with products [22, 23].

Virtual Prototypes can also be used for testing the product design with end users. In this case, they can be used and operated by users with the aim of testing specific product features. The design configurations that can be tested are actually limited and must be predefined by the design team. So, users can say which configuration they prefer, but they cannot ask and try any other configuration outside those ones.

14.2.5.3 iVP: Use-Based: Interaction-centered

It is reasonable expecting that the design of a product that is typically used interactively by users would start by focusing on the interaction, instead of on the overall product. Therefore, before designing the product, the focus should be put on the design of the user interaction. In this case, the design team should work on the design of the product use, independently from how the product itself will be designed. iVP is aimed at supporting this activity [24]. This capability is particularly important for the validation of the initial conceptual design of such products, as those for the consumer market. In fact, in this case, the prototype is used early in the design phase for simulating the behavior and the interaction of something that is meant to be used by a person.

In conclusion, changes in design over the last decades have brought us to the present situation, where the product design starts, being focused on the design of the user experience. Proposed new methods and tools for the evaluation of user interaction to use early in the concept phase are based on iVP, which gives target users the possibility to try and test early simulations and variants of the user experience with new products.

14.3 Consumer–Product Interaction

It has been previously claimed that in the evolution of the products, of the users' expectations, and of the design methods, the attention to those who are the users of the products has grown to the point that the design of consumers' products put the user at the center of the design process. The focus is on the so-called *user-centered design*, which primarily addresses the user experience with a product during its conceptual design. The following sections address some definitions and issues related to consumer products, perceived value for users and salient attributes of these kinds of products.

14.3.1 Consumer Products and Their Perceived Value

Consumer products can be defined as goods that satisfy personal needs. They can be described as any tangible personal property for direct use or consumption and that are used for personal, family or household purposes.

From a production perspective, consumer products are the final result of manufacturing and are what a consumer will see on the store shelf. Household supplies and furniture, consumer electronics, jewelry, children's toys, kitchen supplies and automotive accessories are all examples of consumer goods.

Consumer products are typically produced in volumes. Consequent their scales of production, consumer products represent a major global design and manufacturing sector responsible for the production and supply of diverse goods. It can be reasonably said that consumer goods are universally the most competitive and fastest growing market compared with other products purchased only by industry and businesses.

The perceived value of a product is the personal estimation a consumer makes of it. The consumer's perceived value of a good affects the price that he is willing to pay for it. Generally speaking, consumers are unaware of the true cost of production for the products they buy. Instead, they simply have an internal feeling for how much certain products are worth to them. Thus, in order to obtain successful products and a higher price for their products, producers may pursue marketing strategies to create a higher perceived value for their products [25].

Traditionally, the reasoning about relative perceived value of products has been centered on fundamental concepts as features and functions, and cost. Customers buy products for their features (e.g., a mobile phone with a touch screen) or their specific functionalities (e.g., a mobile phone playing video and music). Similarly, potential customers make purchase decisions considering a product's price. That is, how much a customer thinks that a product will cost him.

Today, these perceptions may not completely reflect reality. Recently, people tend to buy products for the perceived benefits that they deliver. Features and functions, which are the main focus of product design specifications, are sort of container for delivering the benefits that are desired by customers. A product may meet objective performance criteria, which are typically validated by analytical or physical laboratory tests. But a product is successful only if the customers and consumers recognize that the product actually delivers some benefits. And it is also worth noticing that the perception of the customer about the product is central in the purchasing experience.

Formally, the value of a product may be conceptualized as the relationship between the consumer's perceived benefits in relation to the costs of receiving these benefits [25]. It is often expressed as the equation: value = benefits/cost. Value is thus subjective, that is, a function of consumers' estimation.

Today, regarding consumer products, the consumption experience is the true value of a product for consumers. A product purchase can be directed not directly toward a physical product, but rather toward a consumption experience, which consists of both cognitive and emotional activities [26].

Therefore, some of the issues that people consider before placing a value on a product are as follows:

- *Attributes* They include characteristics like size and color. For example, one consumer may prefer a pink mobile phone cover to a black one.
- *Functional benefits* This is what the consumer expects to gain from the product. It concerns all the advantages a product offers as compared to similar product offerings. For example, a product is easy to use, its design meets the expectations of the customer, it is easily available and it has a long lifetime.
- *Emotional benefits* This refers to the feelings evoked in customers while and after buying a product. The quality and reputation of the brand as well as the characteristics of the product play an important role in stirring feelings in customers [4]. The feeling can be contentment, angry, excitement, etc. For example, a lady decides to buy a dress according to how it fits on her body and also for the tactile feedback provided by the fabric.

14.3.2 Salient Attributes of Products

Various attributes, named salient attributes, can be associated with products, which can be classified as follows:

- geometric/shape attributes
- material attributes
- functional attributes
- experience attributes

There are some products mainly containing shape attributes, which are visually salient attributes. These products can be assessed and selected by consumers merely through vision. An example of such product is a vase. When material attributes are part of a product, consumers typically desire to obtain further information using their hands to touch and feel the products, in addition to the visual exploration. A typical example is a dress. Functional attributes are related to what a user can do when using a product and to its performance. For example, some radio stations can be directly connected with our iPod and allow us to listen to our music playlist. Furthermore, there are some attributes of a product, which we name *experience attributes*, that consumers can only assess through actual use, or direct contact, with the products. These are, for example, taste, fit, softness and also ways of using the product through buttons, touch screens.

People explore everyday objects through sensory dimensions, which include vision, touch, hearing, smell and taste. Different types of salient attributes involve different sensory input in regard to consumption experience. For example, a pan can be experienced through vision and touch (to check its appearance and how much it weighs), a bar of soap through smell (to feel its fragrance). But basically, according to some studies, in most cases, people tend to explore objects through two sensory dimensions: vision and touch [27].

Already during a purchase experience, consumers learn about products through both direct and indirect experience [28]. The most effective is the direct experience, which happens through an actual contact with the product. This means touching, holding, handling, manipulating and using the product. Consequently, a major issue for companies is to identify which types of attributes are dominant, which are important to consider when customers make purchase decisions. Therefore, it is important to be able to check the probability that a product can provide an expected experience before customers are confident in making a purchase decision. In order to be effective and relatively inexpensive, this check should be made at the concept design phase, when it is possible to perform the simulation of a consumption experience for consumers, which is assessed prior to the purchase or actual use of the product.

14.4 User Experience Design

User experience design (UXD) is part of the design of a new product [29]. UXD includes the definition of the interaction components, that is, their shapes, material and layout, and the definition of their behavior to users' actions.

UXD refers to the application of user-centered design practices to generate predictive and desirable designs based on the consideration of users' experience with a system. The practice includes the definition of the user interface, graphics, and physical and manual interaction performed in close collaboration with target users.

Two very similar products, with the same look and the same functionality, can deliver experiences of different quality that users like differently [30]. Therefore, the user experience, once designed, needs to be tested with users in order to measure its quality and appreciation.

The experience of a person with a product occurs through direct contact and use. For example, the shape and material that constitute an object are relevant, as well as the modality of use and interaction with it. For this reason, a product should be designed keeping in mind and focusing on the direct and physical interaction.

Part of UXD is the *haptic interaction design* (HID). HID is that phase of product development where one designs the interaction with the product that occurs through touch and manual control [31, 32]. This activity includes the design of the physical interaction components and devices, as well as the design of the modalities for interacting with them, also including the integration with other modalities as vision and sound (also named cross-modal interaction).

The kind of interaction designed depends on the product attributes and on its functionalities and also on the target users of the product. We can design products integrating haptic and touch as a novel interaction modality, which proposes users a new way of doing the usual things. Or we can even design a novel physical interaction that aims at creating new emotional and compelling experiences for the potential future users of the product. The design of haptic interaction devoted to skilled users requires the acquisition of knowledge about the users' skills and dexterity, about the users' objectives in using the product, so as to best exploit the ways these users are used to do things and perform manual tasks.

So, the design team must deal with the design of the experience of a person with a product, so as to optimize his enjoyment, satisfaction and positive emotional response. A major issue is how to check if the designed experience is really satisfactory. This issue is actually difficult to address, since it implies that it is necessary to have a properly functioning product that one can use for testing, before the real final product is available. iVP, based on the simulation of products behavior and use, has demonstrated to be effective for this purpose.

14.5 Interactive Virtual Prototyping

This section introduces the concepts related to *interactive Virtual Prototyping* (iVP), which is proposed as a product concept evaluation methodology for testing the use of products with users, for evaluating the users' experience and ultimately for compiling the list of specifications for product design [20, 33].

Interactive Virtual Prototypes can be used in two ways:

1. for verifying and validating the behavior of a designed product with people (potential customers and consumers) and also
2. for evaluating variations in interaction experiences and identifying the ones that people (potential customers and consumers) like more.

In the first case, the iVP can be used for testing the design solutions that respond to user's needs and requirements, and this is typically done when the design is advanced enough. The second one can be used to test proposed novel interactive solutions with users, when the product has not been yet fully decided and designed.

In both cases, for being beneficial, iVPs should be used in a continuous design-validation loop, where a new interactive modality can be designed, soon after tested with users, and modified if necessary. This close loop requires that the iVP can be easily and quickly modified and adapted to meet the user's preferences.

In order to satisfy all these issues, the iVP should have the following characteristics:

- *Realism and fidelity* The iVP should be perceived by the human senses exactly as a real prototype would be, hiding the complexity of the simulation and of the technology at its basis.
- *Multimodality and multisensory interaction* The iVP should support the same interaction modalities and stimulate the same sensorial channels, as it happens when humans interact with real products.
- *Real-time feedback* The iVP should react to user's actions in real time (from the users' perceptual point of view), with no perceivable delays. This requirement impacts also on the complexity of the product and on the behavioral simulation performances.
- *Parametric* The iVP should be based on functional models describing the behavior of each component to test. The models should be parametric, so that the behavior can be changed until an optimum is reached.
- *Sharable among different users remotely located* Sometimes the testing activities on the same product might be performed in different placed and also in different cultural contexts. Therefore, interactive Virtual Prototypes should be effectively used for this purpose.

14.5.1 iVP Architecture

The architecture of a framework for iVP consists of three main components: the Functional Mock-Up (FMU), consisting of functional models, describing the product behavior and the user interaction model, including a perceptual model for each human sense, which actually implements the user interaction, and the multisensory VR environment for performing and handling the user interaction in a realistic way (Fig. 14.4). The three components are described in the following.

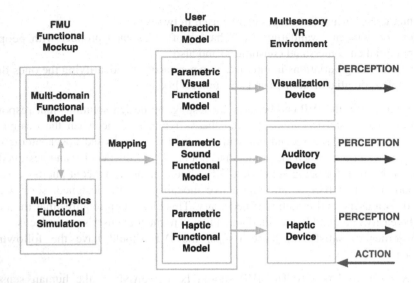

Fig. 14.4 Framework for iVP

14.5.1.1 Functional Mock-Up

The FMU includes functional models and related functional simulation tools. The FMU is based on the definition of several elements, one for each effect that one wants to simulate. A set of parameters is used to describe the behavior of each element, which is simulated by multiphysics functional simulation tools. All these elements are integrated into a unique and comprehensive model, named the multi-domain functional model. As an example, let us consider the FMU built for simulating the door of a refrigerator. In a real refrigerator, there can be several functional elements, each representing an effect: the element simulating the effect of the magnetic attraction of the door when it is closed, the element simulating the effect of the gasket when opening/closing the door, the element related to the sucker-like effect generated by the air contained in the refrigerator, the friction and inertial effects generated when opening/closing the door, etc. These elements are set as parametric, and the parameter values can be changed for setting the final behavior perceived by the user during the interaction, which is performed by the multiphysics simulation tool. FMU is generally a mathematical complex structure, whose computation is time-consuming and typically not running at real time. For this reason, they cannot be used straightforward in interactive applications, and alternative solutions are required for mapping the output of these models into values that can be handled at real time.

14.5.1.2 User Interaction Model

The output of the FMU consists of parameters concerning velocity, acceleration and force related to the simulated effects. These parameters can be mapped into a *user interaction model*, which consists of a set of parametric perceptual models that define the behavior of the product defined at the user perception level. Each model addresses a sensory channel: visual, sound and haptic. These models, including their parameters, are rendered to the user through appropriate VR technologies (3D displays, stereo glasses, haptic devices, auditory devices, etc.) [21]. While visualization and auditory interaction modalities are output modalities, the haptic interaction modality is of type input/output, which can be activated by users' actions, and which reacts accordingly by exerting force-feedback. These models are simplification of the FMUs, which can be used to render the effects at real time. For example, the behavior of the refrigerator door previously considered as an example can be simplified as a spring/damper element described by a force–displacement relation. This parametric model, where the displacement values can be changed, can be rendered to the user as a haptic feedback.

14.5.1.3 Multisensory Virtual Reality Environment

The third component consists of a *Multisensory Virtual Reality Environment* allowing multisensory and multimodal interaction. In the section about the evolution of design tools, it was mentioned that the practice of Virtual Prototyping has been based for years purely on visual representation of products. Actually, iVPs based on the solely visualization of the product are not good for testing the interaction with the product. For example, the study of the use of a door handle is ineffective if performed using a virtual model of the handle. In fact, the user cannot feel the contact with the handle, the mechanical response, etc. Therefore, it is evident that the realistic simulation of the physical contact of the user with the component is important and allows users to feel and evaluate the geometrical shape of the product component, as well as the material properties, as elasticity, rigidity, plasticity, hardness, texture, etc.

For what concerns the physical interaction with the virtual prototype, modern haptic technologies can be used to simulate the physical interaction with a product component [24]. General-purpose haptic devices are not effective for the simulation of the interaction with any object, since they may have limited degrees of freedom, low force-feedback rendering values, or end-effector shapes that do not conform to the component shape [34]. A possible solution to this problem is to develop ad hoc force-feedback systems. Examples are the refrigerator door proposed in [35] and the car door described in [36]. Alternatively, it is possible to use commercial devices where the end effector is replaced or integrated with more appropriate handling tools. This second option, more flexible and general purpose with respect to the first one, has demonstrated to be feasible and effective in various set ups [37]. For example, the end effector of a MOOG HapticMaster device [38] can be replaced with a door handle. An additional benefit is obtained if the handling tool is designed and then produced using a rapid prototyping technique. In fact, it can be easily and relatively quickly re-designed and adapted so as to meet the users' preferences.

14.5.2 iVP and Users Tests

Interactive Virtual Prototyping is a practice used for validating the user interaction with a product, which has been newly designed and is simulated in its various aspects. The validation is done through tests performed with users' groups, during which the designed interaction effects are tested and evaluated. If the effects are considered unsatisfactory or unpleasant, they can be modified.

Two modalities can be used for modifying the product behavior and the interaction effects:

1. Change the parameter values of the FMU. As an example, let us consider the development of the iVP of a knob. The FMU includes the dynamic friction model that describes the knob behavior when it is turned. The FMU is translated into an interaction model, consisting of torque values, which determine the behavior of the knob and which are rendered as haptic feedback by means of a haptic device. The user can ask to modify the reaction to turn. This can be done by modifying the dynamic friction model describing the knob behavior. Once modified, the FMU has to be re-mapped into an updated interaction model.

2. Change the parameter values of the user interaction model. One of the benefits of operating at the level of the interaction model is that the modification of the effects can affect only one component, and the new configuration can be evaluated by considering the overall behavior of the product. Referring to the previous example, the knob dynamic friction model is mapped into a haptic parametric model, consisting of torque values. Besides, it can be added a sound parametric model, consisting of a set of clicks sounds, which is defined only at the interaction level. The effect conveyed to the user is both the force returned and the sound clicks played when the user rotates the knob. The user can ask for making the reaction force of the knob stronger, but without changing the quality and kind of the sound clicks. The modification of the haptic behavior can be done easily and directly by changing the torque values in the interaction model, without affecting the sound model and its output. This is feasible since the interaction models for haptic and sound are independent. The FMU has to be changed according to the changes made in the model at the interaction level.

At the end of the test sessions, the FMU includes all the functional parameters to use for designing the product, which the users tested, assessed and validated. The information included in the FMU in terms of functional model and parameter values can be used straightforward by an engineer as specifications for the definition of the CAD model of the product and its subsystems and components.

14.6 Case Study

The design methods based on the use of iVP presented in the previous sections have been applied and tested for the design and evaluation of the experience of use of some consumer products. Two examples have been selected:

1. FMU of a dishwasher
2. Multisensory interaction model of a refrigerator door

14.6.1 FMU of a Dishwasher

The first case study consists of the simulation of the force returned on the user's hand while opening and closing the door of a dishwasher. The aim of the simulation is to analyze the interaction with the door and identify some new specifications for the re-design of the haptic feedback provided by the interaction with the door of new dishwashers, in accordance with the study results. The case study has been proposed by a company operating in the field of household appliances.

The analysis of the opening effect shows that the haptic feedback is a combination of the initial leaf spring (this is the click effect perceived at the beginning of the opening) and of the inertia of the door. Furthermore, static frictions and compression spring effects complete the overall effect. Traditionally, the door opening/closing force effect is not designed with particular attention to the experiential effects on the users (experience attributes), but it is designed so as to prevent any free movement of the door during its use (functional attributes). For example, when the user opens the door and subsequently releases it, the door does not slam and is gradually braked up. This suggests that the design has focused on the functionality and performance of the product, rather than on the final effect experienced by the users.

On the basis of the door behavioral analysis, it has been defined a model of the door and a simulation of the opening effect. The simulation has been performed using the LMS-AMESim suite [39], which is one-dimensional simulation software tool for multidomain systems. Figure 14.5 illustrates the sketch of the door mechanical system developed using LMS-AMESim where the main parts contributing to the force are highlighted. In particular, it can be seen the contribution of the friction, of the springs and of the leaf spring. Furthermore, the sketch includes a function describing the human force exerted on the door. The one represented is a simulation of the human behavior while opening and closing the door, which is based on the results described in the work by Jain et al. [40], as well as on the results of a set of acquisitions performed on the door of the dishwasher, as described in [41]. Basically, the behavior is represented as a ramp function. Different ramps can be simulated easily through a batch-run simulation, so as to verify the behavior of the door at different accelerations. This is very important in order to understand the efficiency of the door in different opening and closing conditions.

Figure 14.6 graphically shows the results of the behavior of the door obtained by changing the friction constant values.

If we aim at enabling the user to test the behavior of the door through a haptic interface, as, for example, the MOOG HapticMaster [38], we must take into account some important issues. The user's input that has been treated as a ramp with varying peaks in the simulation described so far is actually varying continuously during the opening action. The door system reacts consequently to the user's input. In order to return to the user a realistic feedback in dynamic conditions, we should capture the force exerted by the user's hand continuously and compute the angular position of the door in real time. This is unfeasible at the frequency required by the device to grant a fluid haptic interaction, which is typically of or is higher than 1 kHz.

A way to overcome this limitation consists of simplifying the model that can be tested by the user until real time is obtained. This can be done for example by summing up all the friction contributions so that eventually only one overall friction will be simulated. The same approach can be adopted for springs, dampers, etc. This simplified model is then translated into parametric equations that are rendered through the haptic device. The user can test the force profiles and tune the haptic behavior of the door by asking some modifications that will be translated into parameters of the equations. Finally, the results can be used as input to the LMS-AMESim simulation model, by means of optimization tools.

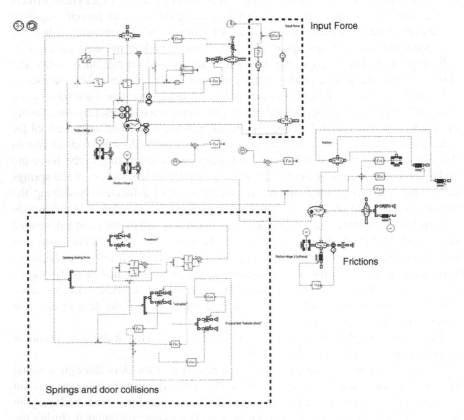

Fig. 14.5 Sketch of the door dishwasher mechanical system

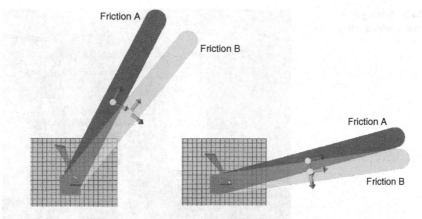

Fig. 14.6 Effects of changing the friction constant values for the door opening

14.6.2 Multisensory Interaction Model of a Refrigerator Door

This second case study consists of designing the experience of a user interacting with the door of a new refrigerator. For what concerns the functional model, the effect of opening the door has been derived from the analysis of the CAD models and from some measurements performed on a real refrigerator door. Specifically, it has been analyzed the door, its shape, users' handling and the door physical behavior. Subsequently, it has been simulated the physical handling of the door. The overall effect has been divided into the following steps:

- initial effect due to the mechanism for keeping the door close;
- subsequent effect due to inertia and friction of the components;
- collisions occurring at the end of the stroke;
- sound made by the moving parts, also depending on the velocity of movement.

The parametric haptic functional model is as such that the effect transmitted to the users is a force returned when they try and open the door. This force is modifiable on user requests until an optimum is obtained. The model of the forces is implemented through a piecewise function, where each subfunction is parametric and can be changed in real time without affecting the other interval subfunctions. The user can ask higher or lower reaction force for each interval; this will be obtained by modifying the parameters. When the test is over, these changes are stored and subsequently mapped into design specifications.

Regarding the haptic device, the initial hypothesis was to use the end effector of a commercially available haptic device for the physical handling simulation. Unfortunately, no haptic device allows us to realistically simulate the shape of any handle. Therefore, it was decided to use an ad hoc physical component for at best representing the interaction experience. This physical component representing the door handling part has been developed using rapid prototyping manufacturing

Fig. 14.7 Testing user experience with a refrigerator door

methods. Then, it has been then mounted on the MOOG HapticMaster system [39], which has been used for rendering the forces (Fig. 14.7).

14.7 Conclusion

In products where the user interaction is one of the main features determining their value, the design of user experience gains importance in view of a successful product. In fact, people often buy products on the basis of the first impression they have about their features gathered during the short use and interaction with them occurring at the selling points. Designing the experience with a new product is not an easy task, especially when the product is radically new and proposes very new interaction modalities. Measuring at what extent a product satisfies the users' needs—spoken or unspoken—is still an open issue.

Therefore, testing the designed experience soon in the product development process becomes crucial for developing a successful product. Actually, testing activities require modalities for presenting the design proposals and variants to the users in a proper and clear way. Prototyping is certainly an effective way for demonstrating a product and its interaction modalities. But if the prototype is physical, it is costly and hardly modifiable. Conversely, the first examples of design experience tests based on interactive virtual prototypes have shown the potentiality and applicability of this very new approach. The effectiveness depends on several factors, including the fidelity of the prototype and of the interaction, the performances, the capability of rendering multimodal and multisensory effects.

In conclusion, on the basis of the initial studies, the use of iVP for designing and testing user experience of consumer products is a course that must be pursued and investigated further. The subject is very interdisciplinary, requiring expertise in many domains as engineering, design, marketing, human factors and neuroscience.

References

1. Goldenberg J, Mazursky D (2002) Creativity in product innovation. Cambridge University Press, Cambridge
2. Bordegoni M (2010) Exploitation of designers and customers' skills and creativity in product design and engineering. In: Fukuda S (ed) Emotional engineering: service development. Springer, London
3. Sweeney JC, Soutar GN (2001) Consumer perceived value: the development of a multiple item scale. J Retail 77(2):203–220
4. Akshay RR, Kent B (1989) The effect of price, brand name, and store name on buyers' perceptions of product quality: an integrative review. Monroe J Mark Res 26(3):351–357
5. Norman DA (1988) The design of everyday things. Doubleday, New York
6. Spence C, Gallace A (2011) Multisensory design: reaching out to touch the consumer. Psychol Mark 28(3):267–308
7. MacLean KE (2009) Putting haptics into the ambience. IEEE Trans Haptics 2(3):123–135
8. Magnusson C, Szymczak D, Brewster S (eds) (2012) Haptic and audio interaction design. Information systems and applications, incl. internet/web, and HCI. Lecture notes in computer science, vol 7468. Springer, Heidelberg, pp 131–140
9. Oberg E (2008) Machinery's handbook. Industrial Press Inc., New York
10. Sakao T, Lindahl M (eds) (2010) Introduction to product/service-system design. Springer, London
11. Fukuda S (2012) Customer productivity: a measure for product and process development with customers. In: Proceedings of ASME IDETC/CIE conference
12. Ulrich KT, Eppinger SD (2004) Product Design and Development. McGraw-Hill, New York
13. Schumpeter JA (1934) The theory of economic development. Harvard University Press, Cambridge
14. Wicker AW (1969) Attitudes versus actions: the relationship of verbal and overt behavioral responses to attitude objects. J Soc Issues 25(4):41–78
15. Loftus E, Wells G (1984) Eyewitness testimony: psychological perspectives. Cambridge University Press, New York
16. Roomba robot (2012) http//www.irobot.com/. Accessed July 2012
17. Segway (2012) http://www.segway.com/. Accessed July 2012
18. The Economist web page (2012) http://www.economist.com/topics/digital-fabrication. Accessed July 2012
19. Doellner G, Kellner P, Tegel O (2000) DMU digital mock-up and rapid prototyping in automotive product development. J Integr Des Process Sci 4(1):55–66 (IOS Press)
20. Bordegoni M (2011) Product virtualization: an effective method for the evaluation of concept design of new products. In: Bordegoni M, Rizzi C (eds) Innovation in product design—from CAD to virtual prototyping. Springer, London, pp 117–142
21. Burdea G, Coiffet P (2003) Virtual reality technology, 2nd edn. Wiley, New Jersey
22. Burdea GC (2000) Haptic feedback for virtual reality. Int J Des Innov Res 2(1):17–29 (Special issue on Virtual Prototyping)
23. Biggs SJ, Srinivasan MA (2002) Haptic interfaces. In: Stanney K (ed) Handbook of virtual environments: design, implementation, and applications. Lawrence Earlbaum, Inc., London, pp 93–116
24. Ferrise F, Bordegoni M, Cugini U (2012) Interactive virtual prototypes for testing the interaction with new products. Comput Aided Des Appl 10(3):515–525
25. Doyle P (2000) Value-based marketing: marketing strategies for corporate growth and shareholder value. Wiley, Chichester
26. Lofman B (1991) Elements of experiential consumption: an exploratory study. In: Holman RH, Solomon MR (eds) Advances in consumer research, 18th edn. Association for Consumer Research, Provo, pp 729–773

242 M. Bordegoni et al.

27. Klatzky RL, Lederman SJ, Matula D (1993) Haptic exploration in the presence of vision. J Exp Psychol Hum Percept Perform 19(4):726–743
28. Millar MG, Millar KU (1996) The effects of direct and indirect experience on affective and cognitive responses and the attitude–behavior relation. J Exp Soc Psychol 32(6):561–579
29. Norman DA (1988) The design of everyday things. Doubleday, New York
30. Buxton B (2007) Sketching user experiences: getting the design right and the right design. Morgan Kaufmann, Interactive Technologies)
31. Benyon D, Turner P, Benyon S (2005) Designing interactive systems: people, activities, contexts, technologies. Addison Wesley, Boston
32. MacLean KE (2008) Haptic interaction design for everyday interfaces. Rev Hum Factors Ergon 4(1):149–194
33. Burdea GC (2000) Haptic feedback for virtual reality. Int J Des Innov Res 2(1):17–29 (Special issue on Virtual Prototyping)
34. Hannaford B, Okamura AM (2008) Haptics. In: Siciliano B, Khatib O (eds) Handbook of robotics. Springer, New York
35. Shin S, Lee I, Lee H, Han G, Hong K, Yim S, Lee J, Park Y, Kang BK, Ryoo HD, Kim DW, Choi S, Chung WK (2012) Haptic simulation of refrigerator door. IEEE haptics symposium (HAPTICS), pp 147–154
36. Strolz M, Groten R, Peer A, Buss M (2011) Development and evaluation of a device for the haptic rendering of rotatory car doors. Ind Electron IEEE Trans Ind Electron 58(8):3133–3140
37. Ferrise F, Bordegoni M, Graziosi S (2012) A method for designing users' experience with industrial products based on a multimodal environment and mixed prototypes. Comput Aided Des Appl 10(3):461–474
38. MOOG (2012) HapticMaster. http://www.moog.com/products/haptics-robotics/. Accessed July 2012
39. LMS (2012) http://www.lmsintl.com/. Accessed July 2012
40. Jain A, Nguyen H, Rath M, Okerman J, Kemp CC (2010) The complex structure of simple devices: a survey of trajectories and forces that open doors and drawers. In: Proceedings of the IEEE RAS/EMBS international conference on biomedical robotics and biomechatronics (BIOROB)
41. Ferrise F, Ambrogio M, Gatti E, Lizaranzu J, Bordegoni M (2011) Virtualization of Industrial Consumer Products for Haptic Interaction Design. ASME Conf Proc 2011(44328):9–18